Lecture Notes in Computer Science   9620

Commenced Publication in 1973
Founding and Former Series Editors:
Gerhard Goos, Juris Hartmanis, and Jan van Leeuwen

Editorial Board

More information about this series at http://www.springer.com/series/8637

Abdelkader Hameurlain · Josef Küng
Roland Wagner (Eds.)

Transactions on Large-Scale Data- and Knowledge-Centered Systems XXV

Springer

Editors-in-Chief

Abdelkader Hameurlain
IRIT, Paul Sabatier University
Toulouse
France

Roland Wagner
FAW, University of Linz
Linz
Austria

Josef Küng
FAW, University of Linz
Linz
Austria

ISSN 0302-9743   ISSN 1611-3349   (electronic)
Lecture Notes in Computer Science
ISBN 978-3-662-49533-9   ISBN 978-3-662-49534-6   (eBook)
DOI 10.1007/978-3-662-49534-6

Library of Congress Control Number: 2015943846

This Springer imprint is published by SpringerNature
The registered company is Springer-Verlag GmbH Berlin Heidelberg

Preface

This volume of *Transactions on Large-Scale Data- and Knowledge-Centered Systems* (TLDKS) contains five fully revised selected regular papers, covering a wide range of hot topics in the field of data and knowledge management systems.

Topics covered include: a framework consisting of two heuristics with slightly different characteristics to compute the action rating of data stores; a theoretical and experimental study of filter-based equijoins in a MapReduce environment; a constraint programming approach that is based on constraint reasoning to study the view selection and data placement problem given a limited amount of resources; a formalization and an approximate algorithm that have been proposed to tackle the problem of source selection and query decomposition in federations of SPARQL endpoints; and a matcher factory that enables the generation of a dedicated schema matcher for a given schema-matching scenario.

We would like to express our great thanks to the editorial board and the external reviewers for thoroughly reviewing the submitted papers and ensuring the high quality of this volume.

Special thanks go to Gabriela Wagner for her availability and her valuable work in the realization of this TLDKS volume.

December 2015

Abdelkader Hameurlain
Josef Küng
Roland Wagner

Editorial Board

Contents

On Expedited Rating of Data Stores

Sumita Barahmand[✉] and Shahram Ghandeharizadeh

Computer Science Department, University of Southern California,
Los Angeles, USA
sumita.barahmand@gmail.com, shahram@dblab.usc.edu

Abstract. To rate a data store is to compute a value that describes the performance of the data store with a database and a workload. A common performance metric of interest is the highest throughput provided by the data store given a pre-specified service level agreement such as 95 % of requests observing a response time faster than 100 ms. This is termed the action rating of the data store. This paper presents a framework consisting of two search techniques with slightly different characteristics to compute the action rating. With both, to expedite the rating process, the framework employs agile data loading techniques and strategies that reduce the duration of conducted experiments. We show these techniques enhance the rating of a data store by one to two orders of magnitude. The rating framework and its optimization techniques are implemented using a social networking benchmark named BG.

1 Introduction

1.1 Motivation

The landscape of data stores has expanded to include SQL, NoSQL, NewSQL, cache augmented, graph databases, and others. A survey of 23 systems is presented in [11] and we are aware of a handful more[1] since that survey. Some data stores provide a tabular representation of data while others offer alternative data models that scale out [12]. Some may sacrifice strict ACID [16] properties and opt for BASE [11] to enhance performance. Independent of a qualitative discussion of these approaches and their merits, a key question is how do these systems compare with one another quantitatively? A single metric that captures both response time and processing capability of a data store is *action rating* [4,5]. It is defined as the highest throughput provided by a data store given a pre-specified service level agreement, SLA. An SLA is a performance agreement between the application developer and the customer. An example SLA may require 95 % of issued requests to observe a response time faster than 100 ms for a pre-specified window of time Δ, say 10 min. The main contribution of this paper is presenting a framework consisting of two search techniques to compute the action rating for a data store. This framework employs strategies to reduce the number of

---

[1] TAO [2], F1 [25], Apache's RavenDB and Jackrabbit, Titan, Oracle NoSQL, Foundation DB, STSdb, EJDB, FatDB, SAP HANA, and CouchBase.

© Springer-Verlag Berlin Heidelberg 2016
A. Hameurlain et al. (Eds.): TLDKS XXV, LNCS 9620, pp. 1–32, 2016.
DOI: 10.1007/978-3-662-49534-6_1

Table 1. BG's SoAR (actions/sec) with two workloads using a database consisting of 100 K members each with 100 friends and 100 resources. The machine hosting the data store is a 64 bit 3.4 GHz Intel Core i7-2600 processor (4 cores hyperthreaded as 8) configured with 16 GB of memory, 1.5 TB of storage, and one Gigabit networking card.

Data Store	100 % View Profile	100 % List Friends
SQL-X	5,714	401
MongoDB	7,699	295
HBase	5,653	214
Neo4j	2,521	112

conducted experiments and utilizes agile data loading techniques to reduce the duration of the experiments.

The action rating reduces the performance of different data stores to one number, simplifying the comparison of different data stores, their data models, and design principles. Workload characteristics that are application specific provide a context for the rating. For example, the BG benchmark [4,5] uses a social graph to generate a workload consisting of interactive social networking actions to evaluate a data store. This is termed the Social Action Rating, SoAR, and can be used to compare different data stores with one another. To illustrate, Table 1 shows the computed SoAR of a document store named MongoDB, an extensible record store named HBase, a graph data store named Neo4j, and an industrial strength Relational Database Management System (RDBMS) named[2] SQL-X. For the imposed workload consisting of a single action that looks up the profile of a member, View Profile, MongoDB is the high performant data store. SQL-X outperforms MongoDB for a workload consisting of 100 % List Friends action even though it joins two tables, see [6] for details.

One may establish the action rating of a data store using either an open or a closed emulation model. With an open emulation model, a benchmarking framework imposes load on a data store by generating a pre-specified number of requests per unit of time, termed arrival rate. This arrival rate, λ, is an average over some period of time and might be modeled as a bursty pattern using a Poisson distribution. With a closed emulation model, the benchmarking framework consists of a fixed number of threads (or processes), T, that issue requests to a data store. Once a thread issues a request, it does not issue another until its pending request is serviced. Moreover, a thread may emulate think time by sleeping for some time between issuing requests. Both the number of threads and the think time control the amount of load imposed on a data store. With both emulation models, one increases system load (either λ or T) until the data store violates the specified SLA. The highest observed number of requests[3] processed per unit of time is the action rating of a data store.

[2] Due to licensing agreement, we cannot disclose the identity of this system.

[3] This is λ with the open emulation model. With the closed emulation model, it is the highest observed throughput, see discussions of Fig. 1.

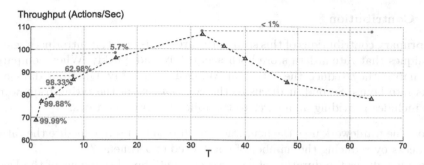

Fig. 1. Throughput as a function of the imposed load (T). Percentage of requests that satisfy the SLA requirement of 100 ms or faster are shown in red.

The action rating is not a simple function of the average service time of a data store for processing a workload. To illustrate, Fig. 1 shows the throughput of SQL-X with a closed emulation model processing a social networking workload generated by BG. On the x-axis, we increase system load by increasing the number of threads T. The y-axis shows the number of actions processed per second by SQL-X. In addition, we show the different percentage of requests that observe a response time faster than 100 ms. With 1 to 4 threads, this percentage is above 95 %. With 8 threads, it drops to 62 %. Beyond 32 threads, the throughput of SQL-X drops as the average service time of the workload starts to increase [30] with less than 1 % of the requests observing a response time faster than 100 ms.

To quantify the action rating of a data store, one must conduct experiments that impose an increasing amount of system load until the highest throughput that satisfies the pre-specified SLA is identified. A naïve technique may perform an exhaustive search of possible system loads by enumerating them starting with the lowest: $\lambda{=}T{=}1$. Alternatively, one may use a search technique such as the Golden Section [31,32] which conducts fewer experiments than naïve and is as accurate. Reducing the number of experiments expedites the rating process because each experiment has a duration in the order of minutes and may have to load a benchmark database on the data store that is being evaluated. The load time is a function of the database size and the target data store. For example, the time to load a modest sized BG database consisting of 100,000 members with 100 friends and 100 resources per member is approximately 3 hours with MongoDB. With MySQL, this time is 8 hours. If these rating techniques conduct hundred experiments to rate MongoDB and fifty experiments to rate MySQL then the time to re-create the database at the beginning of each experiment alone is more than a week.

Ultimately, the time to rate a data store is dependent on the employed search technique that dictates the number of conducted experiments, duration of each experiment, frequency of re-loading the benchmark database in between experiments, the time to load the benchmark database, and the true action rating of the data store. A high performant data store with a million as its rating requires more time to rate than a data store with 100 as its rating.

1.2   Contribution

The primary contributions of this study are three folds. First, it introduces search techniques that rate a data store both accurately and quickly. When compared with naïve, they reduce the rating of MongoDB and MySQL from days and weeks into hours. Second, it discusses the components of the rating framework. This includes providing an answer to the following two questions:

- How the framework uses the heuristic search techniques to expedite the rating process by reducing the number of conducted experiments?
- How it reduces the duration of each experiment? See discussions of the Delta Analyzer in Sect. 3.

Table 2. BG's rating parameters and their definitions. Some of these parameters are used in Sect. 6.

Database parameters	
M	Number of members in the database
ϕ	Number of friends per member
ρ	Number of resources per member
Workload parameters	
O	Total number of sessions emulated by the benchmark
ϵ	Think time between social actions constituting a session
ψ	Inter-arrival time between users emulated by a thread
θ	Exponent of the Zipfian distribution
Service Level Agreement (SLA) parameters	
α	Percentage of requests with response time $\leq \beta$
β	Max response time observed by α requests
τ	Max % of requests that observe unpredictable data
Δ	Min length of time the system must satisfy the SLA
Environmental parameters	
N	Number of BGClients
T	Number of threads
δ	Duration of the rating experiment
r	Climbing factor
Incurred Times	
ζ	Amount of time to create the database for the first time
ν	Amount of time to recreate the database in between experiments
η	Number of rating experiments conducted by BGCoord
ω	Number of times BGCoord loads the database
Υ	Warmup duration
Λ	Total rating duration

Moreover, this framework employs agile data loading techniques to create the benchmark database at the beginning of each experiment. In Sect. 5.2, we show that hardware solutions are not a substitute for a smart strategy to create the benchmark database.

Third, this study details an implementation of the proposed techniques using the BG benchmark. We present experimental results to compare the heuristic search technique when compared with its extensions that employ the agile data loading techniques and techniques that reduce the duration of each experiment. Using Amdahl's law [1], we show these extensions to speedup the heuristic search technique by one to two orders of magnitude, see Table 10.

The rest of his paper is organized as follows. Section 2 presents two variants of our heuristic search technique. In Sects. 3 and 4, we present a technique to reduce the duration of each experiment and three agile data loading techniques, respectively. Section 5 describes an implementation of these techniques using the BG benchmark. An analysis of these techniques including their observed speedup is presented in Sect. 6. We present our related work in Sect. 7. Brief words of conclusion and future research directions are presented in Sects. 8 and 9, respectively.

2 Rating Process

To rate a data store is similar to computing a local maxima. One may compare it to a search space consisting of nodes where each node corresponds to the throughput observed by an experiment with an imposed system load, either T with the closed or λ with the open simulation model. The node(s) with the highest throughput that satisfies the pre-specified SLA identifies the rating of the system. In the following, we focus on a closed emulation model and assume a higher value of T imposes a higher system load on a data store. Section 9 describes extensions in support of an open emulation model.

This section presents two techniques to navigate the search space. The assumptions of both techniques are described in Sect. 2.1. Subsequently, Sect. 2.2 details the two techniques and describes an alternative when the stated assumptions are violated. Section 2.3 compares these two alternatives with one another by quantifying the number of nodes that they visit and their accuracy when rating a data store.

2.1 Assumptions

Our proposed search techniques make the following two assumptions about the behavior of a data store as a function of system load:

1. Throughput of a data store is either a square root or a concave inverse parabola function of the system load, see Fig. 2.a.
2. Average response time of a workload either remains constant or increases as a function of the system load, see Fig. 2.b.

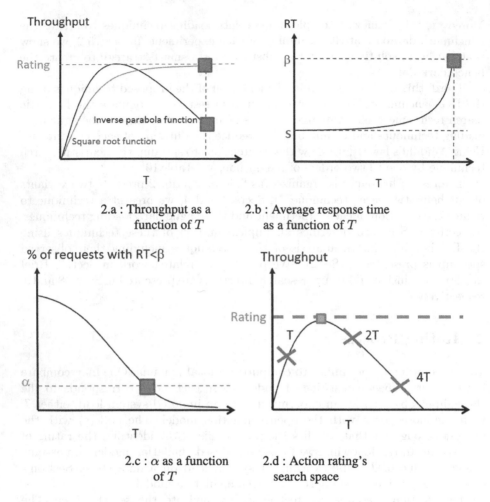

Fig. 2. Assumptions of BG's rating technique.

These are reasonable assumptions that hold true almost always. Below, we formalize both assumptions in greater detail.

Figure 2.b shows the average response time ($\bar{RT}$) of a workload as a function of T. With one thread, $\bar{RT}$ is the average service time ($\bar{S}$) of the system for processing the workload. With a handful of threads, $\bar{RT}$ may remain a constant due to use of multiple cores and sufficient network and disk bandwidth to service requests with no queuing delays. As we increase the number of threads, $\bar{RT}$ may increase due to either (a) an increase in $\bar{S}$ attributed to use of synchronization primitives by the data store that slow it down [9,19], (b) queuing delays attributed to fully utilized server resources where $\bar{RT}=\bar{S}+\bar{Q}$ and $\bar{Q}$ is the average queuing delay, or (c) both. In the absence of (a), the throughput of the data store is a square root function of T else it is an inverse parabola function of T, see Fig. 2.a. In scenario (b), with a closed emulation model where a thread may not issue another request until its pending request is serviced, $\bar{Q}$ is bounded with a fixed number of threads. Moreover, as $\bar{RT}$ increases, the percentage of requests observing an $\bar{RT}$ lower than or equal to β decrease, see Fig. 2.c.

2.2 Search Techniques

The two techniques that employ the assumptions of Sect. 2.1 to rate a data store are as follows. The first technique, named *Golden*, is a search technique assured to compute the rating for the system[4]. The second technique, named *Approximate*, is a heuristic variant that is faster and less accurate. Both techniques realize the search space by conducting experiments where each experiment imposes a fixed load (T) on the system to observe a throughput that may or may not satisfy the pre-specified SLA. Each experiment is a node of the search space. While the search space is potentially infinite, for a well behaved system, it consists of a finite number of experiments defined by a system load (value of T) high enough to cause a resource such as CPU, network, or disk to become 100 % utilized. A fully utilized resource dictates the maximum throughput of the system and imposing a higher load by increasing the value of T (with a closed emulation model) does not increase this observed maximum. A finite value of T limits the number of nodes in the search space.

Both Golden and Approximate navigate the search space by changing the value of T, imposed load. When navigating the search space by conducting experiments, an experiment is *successful* if it satisfies the pre-specified SLA provided by an experimentalist. Otherwise, the experiment has *failed*. Both techniques traverse the search space in two distinct phases: a hill climbing phase and a local search phase. The local search phase differentiates Golden from Approximate. Approximate conducts fewer experiments during this phase and is faster. However, its rating incurs a margin of error and is not as accurate as Golden. Below, we describe the hill climbing phase that is common to both techniques. Subsequently, we describe the local search of Golden and Approximate in turn.

One may implement the hill climbing phase by maintaining the thread count (T_{max}) that results in the maximum observed throughput $(\lambda_{T_{max}})$ among all conducted experiments, i.e., visited nodes of the search space. It starts an experiment using the lowest possible system load, one thread $(T = 1)$ to issue the pre-specified mix of actions. If this experiment fails then the rating process terminates with a rating of zero. Otherwise, it enters the hill climbing phase where it increases the thread count to $T = r \times T$ where r is the hill climbing factor and an input to the technique. (See below for an analysis with different values of r.) It repeats this process until an experiment either fails or observes a throughput lower than $\lambda_{T_{max}}$, establishing an interval for the value of T that yields the rating of the system. Once this interval is identified, the hill climbing phase terminates,

---

[4] We also developed a greedy algorithm of our own named Gauranteed [4]. Gauranteed provides similar performance to Golden except that it visits a higher number of states.

providing the local search space with the identified interval, see lines 15 and 18 of Algorithm 1.

Algorithm 1. Compute SoAR

```
 1: procedure COMPUTE–SOAR
 2:     T_max ← T ← 1
 3:     λ_{T_max} ← 0
 4:     while true do
 5:         (λ_T, SLA_T) ←Conduct an experiment using T
 6:                       to compute λ_T and SLA_T
 7:         if λ_T > λ_{T_max} and SLA experiments are satisfied then
 8:             T_max ← T
 9:             λ_{T_max} ← λ_T
10:             T ← T × r
11:         else
12:             End ← T
13:             if technique is Golden then
14:                 Start ← T/(r×r)
15:                 return Golden–LocalSearch(Start,End)
16:             else
17:                 Start ← T/r
18:                 return Approximate–LocalSearch(Start,End)
```

end

The start of the local search interval is computed differently with Approximate and Golden, see lines 13–18 of Algorithm 1. With Approximate, the starting thread count is $\frac{T}{r}$ and the ending thread count is T and the peak throughput is assumed to reside in the interval $(\frac{T}{r}, T)$. With Golden, the starting thread count is $\frac{T}{r^2}$, the ending thread count is the current T and the peak throughput is assumed to reside in the interval $(\frac{T}{r^2}, T)$. Next, we describe how these local search intervals are navigated by Golden and Approximate.

Golden identifies the peak throughput by maintaining the throughput values for triples of thread counts (nodes) whose distances form a golden ration [31,32]. Next, it successively narrows the range of values inside which the maximum throughput satisfying the SLA requirements is known to exist, see Algorithm 2 for details.

Algorithm 2. Perform Local Search For Golden

```
1: procedure GOLDEN–LOCALSEARCH(Start, End)
2:      GoldenRatio ← (−1+√5)/2
3:      Point1 ← End + GoldenRatio × (Start − End)
4:      Point2 ← Start + GoldenRatio × (End − Start)
5:      while (End − Start > 1) do
6:          (λ_Point1, SLA_Point1) ←Conduct an experiment using Point1
7:                          to compute λ_Point1 and SLA_Point1
8:          (λ_Point2, SLA_Point2) ←Conduct an experiment using Point2
9:                          to compute λ_Point2 and SLA_Point2
10:         if (λ_Point1 > λ_Point2) then
11:             End ← Point2
12:             Point2 ← Point1
13:             Point1 ← End + GoldenRatio × (Start − End)
14:         else
15:             Start ← Point1
16:             Point1 ← Point2
17:             Point2 ← Start + GoldenRatio × (End − Start)
18:     (λ_Start, SLA_Start) ←Conduct an experiment using Start
19:                          to compute λ_Start and SLA_Start
20:     λ_Tmax ← λ_Start
21:     return λ_Tmax
```

end

Approximate navigates the interval identified by the hill climbing phase differently, see Algorithm 3. It treats the start of the interval as the point with the highest observed throughput among all points that have been executed, $\frac{T}{r}$ and its end as the point with the lowest thread count that failed or resulted in a lower throughput, T. It then executes an experiment with the mid-point in this interval. If this experiment succeeds and observes a throughput higher than $\lambda_{T_{max}}$, then the heuristic changes the start of the interval to focus on to this mid-point, does not change the end of the interval (T) and repeats the process. Otherwise, it changes the end point of the interval to be this mid-point, does not change the starting point of the interval and repeats the process until the interval shrinks to consist of one point. It repeats the experiment with this last point as the value of T and compares the observed throughput with $\lambda_{T_{max}}$ to identify the threadcount that maximized the throughput. The heuristic approach is not guaranteed to find the peak throughput (rating) for a system. Its margin of error depends on the behavior of the data store and the climbing factor r. Below, we describe an example to illustrate why Approximate incurs a margin of error.

Algorithm 3. Perform Local Search For Approximate

1: **procedure** APPROXIMATE–LOCALSEARCH(Start,End)
2:     **while** $true$ **do**
3:         $intervalLength \leftarrow End - Start$
4:         **if** $intervalLength < 2$ **then**
5:             **break**
6:         **else**
7:             $T \leftarrow Start + \frac{intervalLength}{2}$
8:             $(\lambda_T, SLA_T) \leftarrow$Conduct an experiment using T
9:                          to compute λ_T and SLA_T
10:            **if** $\lambda_T > \lambda_{T_{max}}$ & SLA requirements satisfied **then**
11:                $Start \leftarrow T$
12:                $T_{max} \leftarrow T$
13:                $\lambda_{T_{max}} \leftarrow \lambda_T$
14:            **else**
15:                $End \leftarrow T$
     return $\lambda_{T_{max}}$

end

Consider a scenario where the experiment succeeds with T threads and increases the thread count to $2T$. With $2T$ the experiment succeeds again and observes a throughput higher than the max throughput observed with T, see Fig. 2.d. Thus, the hill climbing phase increases the thread count to $4T$ (assuming a climbing factor of 2, $r=2$). With $4T$, the experiment produces a throughput lower than the maximum throughput observed with $2T$. This causes the hill climbing phase to terminate and establishes the interval $(2T, 4T)$ for the local search of Approximate. If the peak throughput is in the interval $(T, 2T)$ then Approximate returns 2T as the peak, failing to compute SoAR accurately. The difference between the true peak and 2T is the margin of error observed with Approximate. Golden avoids this error by navigating the interval $(T, 4T)$

Both Golden and Approximate maintain the observed throughput with a given value of T in a hash table. (This is not shown in the Algorithms 1–3.) When exploring points during either the hill climbing phase or local search, an algorithm uses this hash table to detect repeated experiments. It does not repeat them and simply looks up their observed throughput, expediting the rating process significantly.

When comparing throughputs, both Golden and Approximate take some amount of variation into consideration. For example, λ_a is considered higher than λ_b only if it is $\epsilon\%$ higher than it. It is the responsibility of the experimentalist to identify the tolerable variation, ϵ.

With a system that violates the assumptions of Sect. 2.1, both Golden and Approximate may fail to identify system rating. For example, Fig. 3 shows a system where the observed throughput is not an increasing function of system load. In such a case, both techniques may become trapped in a local maxima

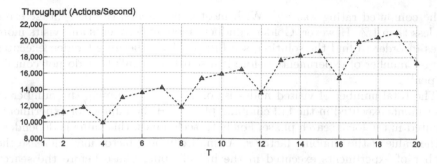

Fig. 3. Behavior of a system violating the rating assumptions.

and fail to identify the peak throughput of the system. A possible approach may use simulated annealing to perform (random) jumps to escape the local maxima. We do not discuss this possibility further as we have not observed a system that violates the stated assumptions.

2.3   Comparison

To compare Golden and Approximate, we use a simple quadratic function, $-aT^2 + bT + c = y$ ($a = 1$ and $b > 1$), to model the throughput of a data store as a function of number of threads issuing requests against it. The vertex of this function is the maximum throughput and is computed by solving the first derivative of the quadratic function: $T = \frac{b}{2}$. Golden and Approximate must compute this value as the rating of the system. We select different values of b and c to model diverse systems whose ratings vary from 100 to 100 million actions per second. We start with a comparison of Golden and Approximate, showing Golden conducts 11 % to 33 % more experiments than Approximate but computes the correct rating at all times. While Approximate is faster it computes an action rating with some margin of error, see discussions of Table 3 below.

Figure 4 shows the number of visited nodes. When the true rating is 100 million actions per second, Golden conducts 69 experiments to compute the value of T that realizes this rating. 35 experiments are repeated from previous iterations with the same value of T. To eliminate these, the algorithm maintains the observed results for the different values of T and performs a look up of the results prior to conducting the experiment. This reduces the number of unique experiments to 34. This is 2.8 times the number of experiments conducted with a system modeled to have a rating of 500 actions per second (which is several orders of magnitude lower than 100 million).

Figure 5 shows the number of unique experiments executed with each of the techniques with Approximate conducting fewer experiments. Figure 6 shows the ideal (expected) rating as well as the computed ratings by the two techniques for the different curves. As shown in Fig. 6, Golden always computes the expected rating for the system. With Approximate, the highest observed percentage error in the computed rating was 8 %. With most experiments, the percentage error was less than 1 %. However, Golden conducts more experiments and visits more nodes in order to find the solution, see Fig. 5. This is because Golden executes a larger number of experiments before it discards intervals that do not contain the peak.

The total number of visited nodes is computed by summing the number of experiments executed in the hill climbing phase and the number of experiments executed in the local search phase. For both techniques, this value is dependent on the value of the climbing factor r. A small climbing factor may increase the number of experiments executed in the hill climbing phase before the search interval is identified. Whereas a large climbing factor may increase the number of experiments executed in the local search phase as it may result in a larger search interval, see Fig. 7.

Table 3 shows a comparison of Approximate with Golden using different climbing factors r. The first column is a known system rating. The next three

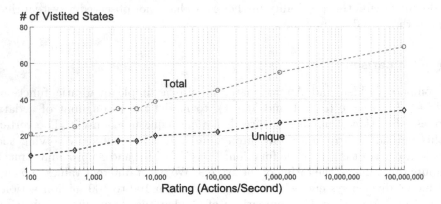

Fig. 4. Number of conducted experiments using Golden.

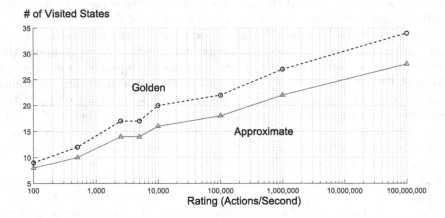

Fig. 5. Number of unique experiments conducted by using Golden and Approximate, $r=2$.

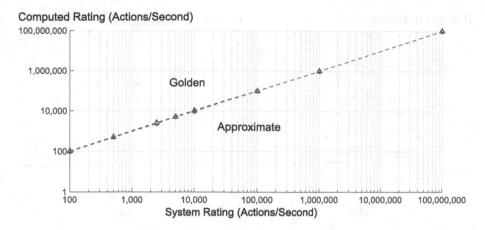

Fig. 6. Rating with Golden and Approximate, $r = 2$.

Table 3. A comparison of Approximate with Golden using different climbing factor r.

System Rating	% Reduction in Experiments			% Error in SoAR		
	$r=2$	$r=4$	$r=10$	$r=2$	$r=4$	$r=10$
100	11.11%	33.33%	25.00%	0.00%	0.00%	0.00%
500	16.67%	25.00%	25.00%	0.00%	0.80%	0.20%
2500	17.65%	25.00%	25.00%	7.84%	7.84%	1.00%
5000	17.65%	20.00%	27.78%	0.08%	0.00%	18.00%
10000	20.00%	25.00%	23.53%	7.84%	1.44%	0.00%
100000	18.18%	25.00%	23.53%	0.40%	0.02%	0.63%
1000000	18.52%	21.74%	21.74%	0.06%	0.06%	0.00%
100000000	17.65%	22.22%	12.00%	0.05%	0.06%	0.00%

columns show the percentage reduction in the number of experiments conducted by Approximate when compared with Golden with 2, 4, and 10 as values of r. The last three columns report the percentage error introduced by Approximate when computing SoAR for the same values of r. These results show Approximate reduces the number of experiments by as much as one third. While the percentage error in the computed SoAR by Approximate is close to zero in most experiments, it was as high as 18% in one instance with $r=10$.

3   Experiment Duration, δ

With the search techniques of Sect. 2, the throughput of each experiment with an imposed system load must be quantified accurately. A key parameter is the duration of each experiment denoted as δ. The ideal duration, value of δ, is

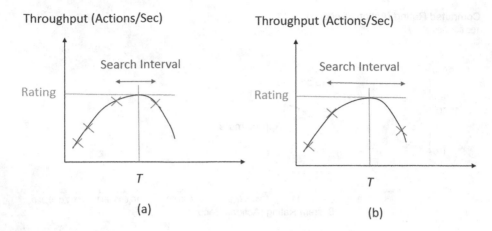

Fig. 7. The impact of the value of climbing factor on the rating, (a) small versus (b) large climbing factor.

both data store and workload dependent. It must be long enough to satisfy two constrains. First, it must generate a mix of requests that corresponds to the specified workload. Here, a high throughput data store shortens the duration of δ because it generates many requests per unit of time to approximate the specified workload quicker than a low throughput data store. Second, the ideal δ value must be long enough to enable a data store to produce a throughput that is permanent[5] with longer δ values. The specified workload and its mix of actions play an important role here. If there is a significant variation in the service time of the data store for the specified actions then the value of δ must be longer to capture this variation. This section describes a technique named Delta Analyzer, DA, to compute the ideal δ.

DA consists of a core timing component, DA-Core, that is invoked repeatedly. (DA-Core might be realized using the component that is used by the search techniques of Sect. 2 to compute the throughput of a node, see Sect. 5 for details.) The input of DA-Core is the user defined workload and the amount of load that it must impose, T. Its output is the value of δ, the observed resource utilization, and the mix of requests issued. DA-Core generates the input workload and system load for t time units repeatedly, doubling the value of t each time. The starting value of t is configurable and might be 1 second by default. DA-Core terminates when two conditions hold true. First, the mix of requests issued for the last q experiments is approximately the same as the input workload. Second, the observed throughput and resource utilization does not change beyond a certain threshold for these q iterations. These last q iterations include t, $2t$,..., $2^{q-1}t$. The value of q is a configurable parameter of DA-Core. The value of t is the ideal δ value for the specified workload and system load T.

---

[5] One may define permanent using a tolerable threshold, say $\pm 10\%$, on the variation in the observed throughput.

DA invokes DA-Core with a low system load, $T=1$, to quantify the value of t and utilization of resources. It then identifies the resource with the highest utilization and uses its reciprocal to estimate the multiplicative increase ρ in the value of T to cause this resource to become fully utilized. It invokes DA-Core with the new value of ρT to establish a new value for δ. This process repeats until a resource becomes fully utilized. The definition of full utilization of a resource is configurable and might be set at 80 % usage. Once this termination condition is reached, DA terminates and reports the value of $\delta = \frac{2^{q-1}t}{2^{q}-1} = t$ as the ideal experiment duration.

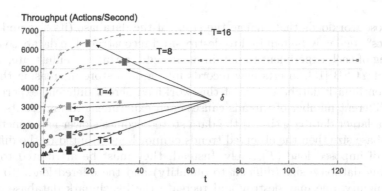

Fig. 8. DA with BG and a social graph of 10,000 members, 100 friends per member, and 100 resources per member. Workload consists of 0.1 % write actions.

To illustrate, Fig. 8 shows the behavior of DA with $q=3$ and 5 % tolerable change in the observed throughput and system utilization with BG using a workload that has a low frequency (0.1 %) of write actions [6]. The x-axis of this figure shows the different values of t. The y-axis is the observed throughput with different values of t employed by DA-Core, starting with $t=1$. The curves show the different system loads considered by DA, i.e., invocation of DA-Core with $T=1$, 2, 4, 8, and 16 threads. When DA invokes DA-Core with a low system load, $T=1$, the termination condition is satisfied with $t=32$ s. In this experiment, the ideal δ is 8 s, i.e., $\frac{t=32}{2^{(3-1)}}$. The network utilization is at 11 % and higher than both CPU and disk. As DA doubles the system load by changing the value of T to 2, 4, 8, and 16, the network utilization increases sub-linearly to 22 %, 44 %, 75 %, and 91 %. It is interesting to note that the throughput observed with 8 and 16 threads is almost identical with small values of t, i.e., 1 and 2 s. The observed throughput is higher with $T=16$ threads and values of t higher than 4 s. The ideal delta is 16 s and twice that observed with a low system load. One may configure DA to report the maximum of the candidate delta values or the delta value observed with the highest system load as the duration of experiments, δ, conducted by the heuristic search technique.

Table 4. Time to load (minutes) a BG social graph with 10,000 and 100,000 members into three data stores.

Data Store	10K Members	100K Members
MongoDB	2	157
SQL-X	7	153.5
MySQL	15	2509

4   Agile Data Loading

With those workloads that change the state of the database (its size and characteristics[6]) one may be required to destroy and reconstruct the database at the beginning of each experiment to obtain meaningful ratings. To illustrate, Workload D of YCSB [13], inserts new records into a data store, increasing the size of the benchmark database. Use of this workload across different experiments with a different number of threads causes each experiment to impose its workload on a larger database size. If the data store becomes slower as a function of the database size then the observed trends cannot be attributed to the different amount of imposed load (T) solely. Instead, they must be attributed to both a changing database size (difficult to quantify) and the offered load. To avoid this ambiguity, one may destroy and recreate the benchmark database at the beginning of each experiment.

This repeated destruction and creation of the same database may constitute a significant portion of the rating process, see Table 4. As an example, the time to load a modest sized BG database consisting of 10,000 members with 100 friends and 100 resources per member is 2 min with MongoDB. With an industrial strength relational database management system (RDBMS) using the same hardware platform, this time increases to 7 min. With MySQL, this time is 15 min [6]. If the rating of a data store conducts 10 experiments, the time attributed to loading the data store is ten times the reported numbers, motivating the introduction of agile data load techniques to expedite the rating mechanism.

This section presents three agile data loading techniques. The first technique, named *RepairDB*, restores the database to its original state prior to the start of an experiment. One may implement RepairDB using either point-in-time recovery mechanism of a data store or techniques that are agnostic to a data store. Section 5.1 presents an implementation of the latter.

The second technique, named Database Image Loading, *DBIL*, relies on the capability of a data store to create a disk image of the database. DBIL uses this

---

[6] Cardinality of a many-to-many relationship such as the number of friends per member of a social networking site.

image repeatedly across different experiments[7]. Depending on the percentage of writes and the data store characteristics, RepairDB may be slower than DBIL. When the percentage of write actions is very low, RepairDB might be faster than DBIL.

The third technique, named *LoadFree*, does not load the database in between experiments. Instead, it requires the benchmarking framework to maintain the state of the database in its memory across different experiments. In order to use LoadFree, the workload and its target data store must satisfy several conditions. First, the workload must be symmetric: It must issue write actions that negate one another in the long run. For example, with BG, a symmetric workload issues Thaw Friendship (TF) action as frequently as Invite Friend (IF) and Accept Friend Request (AFR). The TF action negates the other two actions across repeated experiments. This prevents both an increased database size and the possible depletion of the benchmark database from friendship relationships to thaw. See Sect. 5.3 for other conditions that constrain the use of LoadFree.

In scenarios where LoadFree cannot be used for the entire rating of a data store, it might be possible to use LoadFree in several experiments and restore the database using either DBIL or RepairDB. The benchmarking framework may use this *hybrid* approach until it rates its target data store. Section 6 shows the hybrid approaches provide a factor of five to twelve speedup in rating a data store.

An implementation of these techniques in BG is detailed in Sect. 5. Section 6 presents an evaluation of these techniques.

5   An Implementation

Figure 9 shows the software architecture of an implementation of the concepts presented in the previous three sections using the BG benchmark [5]. BG is a scalable benchmark that employs a shared-nothing hardware platform to generate sufficient requests to rate high throughput data stores. Its components include N BG Clients (BClient), one BG Coordinator (BGCoord), and one Delta Analyzer (DA). We describe these in turn. Subsequently, Sects. 5.1-5.3 describe an implementation of the three agile data loading techniques.

BGCoord implements the search techniques of Sect. 2 that conduct repeated experiments by issuing commands to BGClients to conduct experiments to compute the SoAR of a data store. One of BGCoord's inputs is the duration of each experiment computed using the delta analyzer (DA) of Sect. 3. Each BGClient instance consists of a workload generator and an implementation of BG's eleven social networking actions specific to a data store.

To be portable to different operating systems such as Linux and Windows, BG is implemented using the Java programming language. The different components communicate using message passing (instead of operating system specific

---

[7] One may implement a similar loading technique named Virtual Machine Image Loading (VMIL) by creating virtual machines with a clean copy of the database. The resulting VM image is used repeatedly across different experiments. When compared with DBIL, the VM image is larger than just the database image and may require a longer time to copy.

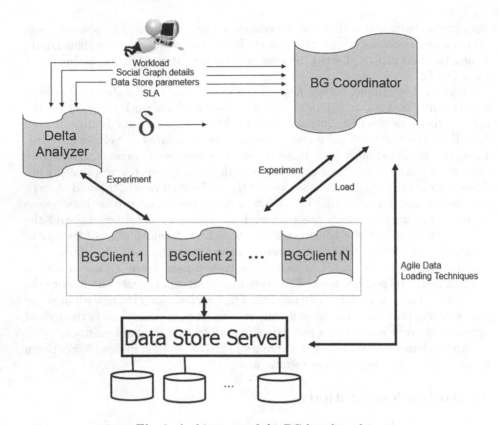

Fig. 9. Architecture of the BG benchmark.

remote invocations) to conduct different rounds of experiments. This is realized using a BGListener (not shown in Fig. 9) that is co-located with each BGClient instance. BGCoord and DA communicate with the BGListener using message passing. BGListener communicates with a spawned BGClient directly. There is one BGListener per BGClient.

Currently, the DA is separated from the BGCoord. Its input include the workload, details of the social graph, and the data store parameters[8]. A workload consists of a mix of eleven social networking actions and the degree of skew for referencing different data items. DA conducts experiments by imposing a different amount of load on a data store using the BGClients. Each BGClient is multi-threaded and the number of threads dictates how much load it imposes on a data store. Each BGClient gathers its observed throughput of a data store along with the utilization of the resources of a server hosting the data store. DA

---

[8] The data store parameters are those required to connect to the database such as the connection URL, database name, data store username and password and all other data store specific parameters such as MongoDB's write concern and read preference.

uses this information to computer δ, the duration of experiments to be conducted by BGCoord, per discussions of Sect. 3.

BGCoord inputs the workload, details of the social graph and the data store, the user specified SLA and δ to rate the data store. It implements the agile data loading techniques of Sect. 4. The software architecture of Fig. 9 is general purpose. One may adapt it for use with YCSB [13] and YCSB++ [24] by specifying a high tolerable reponse time, i.e., max integer, for the input SLA. In addition, the core classes of both YCSB and YCSB++ should be extended to (a) gather resource utilization of servers hosting the data store, and (b) communicate the gathered resource utilization and the observed throughput to BGCoord.

5.1 Repair Database

Repair Database, *RepairDB*, marks the start of an experiment (T_{Start}). At the end of the experiment, it may employ the point-in-time recovery [20,21] mechanism of the data store to restore the state of the database to its state at T_{Start}. This enables the rating mechanism to conduct the next experiment as though the previous benchmark database was destroyed and a new one was created. It is appropriate for use with workloads consisting of infrequent write actions. It expedites the rating process as long as the time to restore the database is faster than destroying and re-creating the same database.

With those data stores that do not provide a point-in-time recovery mechanism, the benchmarking framework may implement RepairDB. Below, we focus on BG and describe two alternative implementations of RepairDB. Subsequently, we extend the discussion to YCSB and YCSB++.

The write actions of BG impact the friendship relationships between the members and post comments on resources. BG generates log records for these actions in order to detect the amount of unpredictable [5] data during its validation phase at the end of an experiment. One may implement point-in-time recovery by using these log records (during validation phase) to restore the state of the database to the beginning of the experiment.

Alternatively, BG may simply drop existing friendships and posted comments and recreate friendships. When compared with creating the entire database, this eliminates reconstructing members and their resources at the beginning of each experiment. The amount of improvement is a function of the number of friendships per member as the time to recreate friendship starts to dominate the database load time. Table 5 shows RepairDB improves the load time of MongoDB[9] by at least a factor of 2 with 1000 friends per member. This speedup is higher with fewer friends per member as RepairDB is rendered faster.

BG's implementation of RepairDB must consider two important details. First, it must prevent race conditions between multiple BGClients. For example, with an SQL solution, one may implement RepairDB by requiring BGClients to drop tables. With multiple BGClients, one succeeds while others encounter exceptions. Moreover, if one BGClient creates friendships prior to another

---

[9] The factor of improvement with MySQL is 3.

Table 5. Factor of improvement in MongoDB's load times with RepairDB compared with re-creating the database, $M=100K$, $\rho=100$.

No. of friends per member (ϕ)	Speedup factor
10	12
100	7
1000	2

BGClient dropping tables then the resulting database will be wrong. We prevent undesirable race conditions by requiring BGCoord to invoke only one BGClient to destroy the existing friendships and comments.

Second, RepairDB's creation of friendships must consider the number (N) of BGClients used to create the self contained communities. Within each BGClient [5], the number of threads (T_{load}) used to generate friendships simultaneously is also important. To address this, we implement BGCoord to maintain the original values of N and T_{load} and to re-use them across the different experiments.

Extensions of YCSB and YCSB++ to implement RepairDB is trivial as they consist of one table. This implementation may use either the point-in-time recovery mechanism of a data store or generate log records similar to BG.

5.2   Database Image Loading

Various data stores provide specialized interfaces to create a "disk image" of the database [23]. Ideally, the data store should provide a high-throughput external tool [24] that the benchmarking framework employs to generate the disk image. Our target data stores (MongoDB, MySQL, an industrial strength RDBMS named SQL-X) do not provide such a tool. Hence, our proposed technique first populates the data store with benchmark database and then generates its disk image. This produces one or more files (in one or more folders) stored in a file system. A new experiment starts by shutting down the data store, copying the files as the database for the data store, and restarting the data store. This technique is termed Database Image Loading, *DBIL*. In our experiments, it improved the load time of MongoDB with 1 million members with 100 friends and 100 resources per member by more than a factor of 400. Figure 10 compares the amount of time it takes to load a social graph consisting of 100,000 members using DBIL with using BG to construct the database for a data store that stores its data on disk, an MLC SSD and a virtual disk[10]. All three are slower than the DBIL technique.

The reason copying an image of a database using DBIL is faster than constructing the social graph using BG is because it does a sequential read and write

---

[10] With disk and the MLC SSD the disk on the node hosting the data store becomes the bottleneck. With the virtual disk, the CPU of the node hosting the data store becomes the bottleneck as now all the data is written to the memory. DBIL is faster than this approach as it eliminates the overhead of locking and synchronization on the data store.

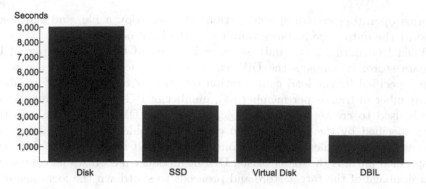

Fig. 10. Time to load a 100 K member social graph with 3 different hardware configurations and an agile data loading technique named DBIL.

of a file. BG's construction of the social graph is slower because it generates users and friendships dynamically[11]. This may cause a data store to read and write the same page (corresponding to a user) many times in order to update a piece of information (a user's JSON object) repeatedly (modify friends). In addition, it also must construct index structure that is time consuming[12].

With DBIL, the load time depends on how quickly the system copies the files pertaining to the database. One may expedite this process using multiple disks, a RAID disk subsystem, a RAM disk or even flash memory. We defer this analysis to future work. Instead, in the following, we assume a single disk and focus on software changes to implement DBIL using BG.

Our implementation of DBIL utilizes a disk image when it is available. Otherwise, it first creates the database using the required (evaluator provided) methods[13]. Subsequently, it creates the disk image of the database for future use. Its implementation details are specific to a data store. Below, we present the general framework. For illustration purposes, we describe how this framework is instantiated in the context of MongoDB. At the time of this writing, an implementation of the general framework is available with MongoDB, MySQL and SQL-X.

We implemented DBIL by extending BGCoord and introducing a new slave component that runs on each server node (shard) hosting an instance of the data store. (The BGClient and BGListener are left unchanged.) The new component is named *DBImageLoader* and communicates with BGCoord using sockets. It

---

[11] Constructing the social graph using BG without actually issuing calls to the data store takes less than a second showing that BG does not impose any additional overhead while loading the social graph into the data store.

[12] In addition, with MongoDB, there is also the overhead of locking and synchronization on the data store.

[13] With BG, the methods are insertEntity and createFriendship. With YCSB, this method is insert.

performs operating system specific actions such as copy a file, and shutdown and start the data store instance running on the local node.

When BGCoord loads a data store, it is aware of the nodes employed by the data store. It contacts the DBImageLoader of each node with the parameters specified by the load configuration file such as the number of members (M), number of friends per member (ϕ), number of BGClients (N), number of threads used to create the image (T_{Load}), etc. The DBImageLoader uses the values specified by the parameters to construct a folder name containing the different folders and files that correspond to a shard. It looks up this folder in a pre-specified path. If the folder exists, DBImageLoader recognizes its content as the disk image of the target store and proceeds to shutdown the local instance of the data store, copy the contents of the specified folder into the appropriate directory of the data store, and restarts the data store instance. With a sharded data store, the order in which the data store instances are populated and started may be important. It is the responsibility of the programmer to specify the correct order by implementing the "MultiShardLoad" method of BGCoord. This method issues a sequence of actions to the DBImageLoader of each server to copy the appropriate disk images for each server and start the data store server.

As an example, a sharded MongoDB instance consists of one or more Configuration Servers, and several Mongos and Mongod instances [22]. The Configuration Servers maintain the metadata (sharding and replication information) used by the Mongos instances to route queries and perform write operations. It is important to start the Configuration Servers prior to Mongos instances. It is also important to start the shards (Mongod instances) before attaching them to the data store cluster. The programmer specifies this sequence of actions by implementing "MultiShardStart" and "MultiShardStop" methods of BGCoord.

5.3   Load Free

With Load Free, the rating framework uses the same database across different experiments as long as the *correctness* of each experiment is preserved. Below, we define correctness. Subsequently, we describe extensions of the BG framework to implement LoadFree.

Correctness of an experiment is defined by the following three criteria. First, the mix of actions performed by an experiment must match the mix specified by its workload. In particular, it is unacceptable for an issued action to become a no operation due to repeated use of the benchmark database. For example, with both YCSB and YCSB++, a delete operation must reference a record that exists in the database. It is unacceptable for an experiment to delete a record that was deleted in a previous experiment. A similar example with BG is when a database is created with 100 friends per member and the target workload issues Thaw Friendship (TF) more frequently than creating friendships (combination of Invite Friend and Accept Friend Request). This may cause BG to run out of the available friendships across several experiments using LoadFree. Once each member has zero friends, BG stops issuing TF actions as there exist no friendships to be thawed. This may introduce noise by causing the performance results obtained in one experiment to deviate from their true value. To prevent this, the workload should be symmetric such that the write actions negate one another. Moreover, the benchmarking framework must maintain sufficient state across different experiments to issue operations for the correct records.

Second, repeated use of the benchmark database should not cause the actions issued by an experiment to fail. As an example, workloads D and E of YCSB insert a record with a primary key in the database. It is acceptable for an insert to fail due to internal logical errors in the data store such as deadlocks. However, failure of the insert because a row with the same key exists is not acceptable. It is caused by repeated use of the benchmark database. Such failures pollute the response times observed from a data store as they do not perform the useful work (insert a record) intended by YCSB. To use LoadFree, the uniqueness of the primary key must be preserved across different experiments using the same database. One way to realize this is to require the core classes of YCSB to maintain sufficient state information across different experiments to insert unique records in each experiment.

Third, the database of one experiment should not impact the performance metrics computed by a subsequent experiment. In Sect. 4, we gave an example with YCSB and the database size impacting the observed performance. As another example, consider BG and its metric to quantify the amount of unpredictable reads. This metric pertains to read actions that observe either stale, inconsistent, or wrong data. For example, the design of a cache augmented data store may incur dirty reads [17] or suffer from race conditions that leave the cache and the database in an inconsistent state [15], a data store may employ an eventual consistency [26,27] technique that produces either stale or inconsistent data for some time [24], and others. Once unpredictable data is observed, the in-memory state of database maintained by BG is no longer consistent with the state of the database maintained by the data store. This prevents BG from accurately quantifying the amount of stale data in a subsequent experiment. Hence, once unpredictable data is observed in one experiment, BG may not use LoadFree in a subsequent experiment. It must employ either DBIL or RepairDB to recreate the database prior to conducting additional experiments.

LoadFree is very effective in expediting the rating process (see Sect. 6) as it eliminates the load time between experiments. One may violate the above three aforementioned criterion and still be able to use LoadFree for a BG workload. For example, a workload might be asymmetric by issuing Post Comment on a Resource (PCR) but not issuing Delete Comment from a Resource (DCR). Even though the workload is asymmetric and causes the size of the database to grow, if the data store does not slow down with a growing number of comments (due to use of index structures), one might be able to use LoadFree, see Sect. 6. In the following, we detail BG's implementation of LoadFree.

To implement LoadFree, we extend each BGClient to execute either in *one time* or *repeated* mode. With the former, BGListener starts the BGClient and the BGClient terminates once it has either executed for a pre-specified[14] amount

---

[14] Described by the workload parameters.

of time or has issued a pre-specified number of requests [13, 24]. With the latter, once BGListener starts the BGClient, the BGClient does not terminate and maintains the state of its in-memory data structures that describe the state of the database. The BGListener relays commands issued by the BGCoord to the BGClient using sockets.

We extend BGCoord to issue the following additional[15] commands to a BGClient (via BGListener): reset and shutdown. BGCoord issues the reset command when it detects a violation of the three aforementioned criteria for using LoadFree. The shutdown command is issued once BGCoord has completed the rating of a data store and has no additional experiments to run using the current database.

In between experiments identified by Execute On Experiment (EOE) commands issued by BGCoord, BGClient maintains the state of its in-memory data structures. These structures maintain the pending and confirmed friendship relationships between members along with the comments posted on resources owned by members. When an experiment completes, the state of these data structures is used to populate the data structures corresponding to the initial database state for the next experiment. BGClient maintains both initial and final database state to issue valid actions (e.g., Member A should not extend a friendship invitation to Member B if they are already friends) and quantify the amount of unpredictable data at the end of each experiment, see [5] for details.

6   Evaluation

This section quantifies the speedup observed with the 3 proposed loading techniques and the DA using a fixed workload. With the data loading techniques, we consider two hybrids: (1) LoadFree with DBIL and (2) LoadFree with RepairDB. These capture scenarios where one applies LoadFree for some of the experiments and reloads the database in between. With the DA, we focus both on Simple[16] and agile data loading techniques to quantify the observed speedup.

In the following, we start with an analytical model that describes the total time required by the heuristic search techniques to rate a data store. Next, we describe how this model is instantiated by the data loading techniques. Subsequently, we describe how the ideal δ impacts the overall rating duration. We conclude by presenting the observed enhancements and quantifying the observed speedup relative to not using the proposed techniques for three different single node data stores[17]. Each node running a data store consists of an Intel i7-4770

---

[15] Prior commands issued using BGListener include: create schema, load database and create index, Execute One Experiment (EOE), construct friendship and drop updates. The EOE command is accompanied by the number of threads and causes BG to conduct an experiment to measure the throughput of the data store for its specified workload (by BGCoord). The last two commands implement RepairDB.

[16] We refer to constructing the database by issuing queries against it as the Simple loading technique.

[17] SQL-X, MongoDB 2.0.6 and MySQL 5.5.

CPU (quad core, 3.8 GHz), 16 Gigabytes of memory, a 1 Terabyte disk drive, and 1 network interface card.

Table 6. BG's rating of MongoDB (minutes) with 1 BGClient for a fixed workload, ϕ=100, ρ=100, Υ =1 min, δ=3 min, Δ=10 min, and η=11. The hybrid techniques used either DBIL or RepairDB for approximately 25 % of experiments.

M	Action	DBIL	RepairDB	LoadFree	LoadFree + DBIL	LoadFree + RepairDB
100K	ζ	165	157	157	165	157
	ν	8	26	0	1.9	6.4
	Λ	308	498	212	242	284
500K	ζ	361	351	351	361	351
	ν	10	165	0	2.5	41.2
	Λ	526	2221	406	444	860
1000K	ζ	14804	14773	14773	14804	14773
	ν	31	588	0	7.75	147
	Λ	15200	21296	14828	14887	16445

6.1 Analytical Model

With BG, the time required to rate a data store depends on:

- The very first time to create the database schema and populate it with data. This can be done either by using BGClients to load BG's database or by using high throughput tools (such as the Bulkload of YCSB++ [24]) that convert BG's database to an on-disk native format of a data store. We let ζ denote the duration of this operation. With DBIL, ζ is incurred when there exists no disk image for the target database specified by the workload parameters M, P, ϕ, ι, ϱ and ρ, and environmental parameter N and others. In this case, the value of ζ with DBIL is higher than RepairDB because, in addition to creating the database, it must also create its disk image for future use, see Table 6.
- The time to recreate the database in between rating experiments, ν. With DBIL and RepairDB, ν should be a value less than ζ. Without these techniques, ν equals ζ, see below.
- The duration of each rating experiment, δ.
- The warmup duration required for each round of experiment, Υ. This duration may be determined either by experience or using DA. One may use DA to compute the amount of time it takes for a system to warm up.
- Total number of rating experiments conducted by the heuristic search techniques, η.
- Total number of times BGCoord loads the database, ω. This might be different than η with LoadFree and hybrid techniques that use a combination of LoadFree with the other two techniques.

– The duration of the final rating round per the pre-specified SLA, Δ.

The total rating duration is:

$$\Lambda = \zeta + (\omega \times \nu) + (\eta \times (\Upsilon + \delta)) + (\Upsilon + \Delta) \tag{1}$$

With LoadFree, ω equals zero. The value of ω is greater than zero with a hybrid technique that combines LoadFree with either DBIL or RepairDB. The value of ν differentiates between DBIL and RepairDB, see Table 6. Its value is zero with LoadFree[18].

By setting ν equal to ζ, Eq. 1 models a simple use of BG's heuristic search technique that does not employ the agile data loading techniques described in this chapter. This technique would require 1939 min (1 day and eight hours) to rate MongoDB with 100 K members. The third row of Table 6 shows this time is reduced to a few hours with the proposed loading techniques. This is primarily due to considerable improvement in load times, see the first two rows of Table 6. Note that the initial load time (ζ) with DBIL is longer because it must both load the database and construct the disk image of the database. The last six rows of Table 6 show the observed trends continue to hold true with databases consisting of 500 K and 1 million members. In addition, if $\delta < \Delta$ is used as the duration of each rating experiment, then the overall duration for of the rating process will improve.

An obvious question is the impact of the discussed techniques while leaving other pieces alone relative to the simple use of the heuristic techniques ($\nu=\zeta$) and when $\delta = \Delta$? Amdahl's Law [1] provides the following answer:

$$S = \frac{1}{(1 - f) + f/k} \tag{2}$$

where S is the observed speedup, f is the fraction of work in the faster mode, and k is speedup while in faster mode. The next two paragraphs will describe how f and k are computed for various agile data loading techniques and the ideal δ computed by DA.

Focusing on the data loading techniques alone, the fraction of work done in the faster mode is computed as $f = \frac{\omega \times \zeta}{\Lambda}$, and the speedup while in faster mode is computed using $k = \frac{\zeta}{\nu}$. With LoadFree, ν is zero, causing k to become infinite. In this case, we compute speedup using a large integer value (maximum integer value) for k because S levels off with very large k values. In [7], we show the value of S to level off with the value of k in the order of hundred thousand.

When only changing the duration of rating experiments from Δ to ideal δ, the fraction of work done in the faster mode is computed as $f = \frac{\eta \times (\Upsilon + \Delta)}{\Lambda}$. Speedup while in faster mode is computed as $k = \frac{(\Upsilon + \Delta)}{(\Upsilon + \delta)}$.

When we use both an agile data loading technique and the ideal δ computed by the DA in our rating experiments, the following are used to compute the overall speedup compared to the simple use of heuristic search without the use of agile data loading techniques: $f = \frac{(\omega \times \zeta) + (\eta \times (\Upsilon + \Delta))}{\Lambda}$ and $k = \frac{(\omega \times \zeta) + (\eta \times (\Upsilon + \Delta))}{(\omega \times \nu) + (\eta \times (\Upsilon + \delta))}$.

---

[18] With LoadFree, a value of ν higher than zero is irrelevant as ω equals zero.

Table 7. Observed speedup (S) when rating MongoDB using agile loading techniques.

M	DBIL	RepairDB	LoadFree	LoadFree + DBIL	LoadFree + RepairDB
100K	6.5	3.9	9.1	8.3	6.8
500K	8.3	1.9	10.5	9.8	5
1000K	11.7	8.3	12	11.9	10.8

Table 8. BG's rating (minutes) of MongoDB, MySQL and SQL-X with 1 BGClient for a fixed workload, M=100K, ϕ=100, ρ=100, ω=11, Υ=1 min, δ=3 min, Δ=10 min, and η=11. The hybrid techniques used either DBIL or RepairDB for approximately 25 % of the experiments.

Data Store	Action	DBIL	RepairDB	LoadFree	LoadFree+DBIL	LoadFree+RepairDB
MongoDB	ζ	165	157	157	165	157
	ν	8	26	0	1.9	6.4
	Λ	308	498	212	242	284
MySQL	ζ	2514	2509	2509	2514	2509
	ν	4.7	1206	0	1.2	302
	Λ	2620	15830	2564	2582	5881
SQL-X	ζ	158.5	153.5	153.5	158.5	153.5
	ν	5	30	0	1.3	7.5
	Λ	274	544	214	232	296

6.2 Speedup with Loading Techniques

Table 7 shows the observed speedup (S) for the experiments reported in Table 6. LoadFree provides the highest speedup followed by DBIL and RepairDB. The hybrid techniques follow the same trend with DBIL outperforming RepairDB. Speedups reported in Table 7 are modest when compared with the factor of improvement observed in database load time between the very first and subsequent load times, compare the first two rows (ζ and ν) of Table 6. These results suggest the following: Using the proposed techniques, we must enhance the performance of other components of BG to expedite its overall rating duration. (It is impossible to do better than a zero load time of LoadFree.) A strong candidate is the duration of each experiment (δ) conducted by BG. Another is to reduce the number of conducted experiments by enhancing the heuristic search techniques.

Reported trends with MongoDB hold true with both MySQL and an industrial strength RDBMS named[19] SQL-X. The time to load these data stores and rate them with 100 K member database is shown in Table 8. For all three data stores Load Free provides the highest speedup followed by DBIL and RepairDB. And the hybrid techniques follow the same trend. While SQL-X provides comparable response time to MongoDB, MySQL is significantly slower than the other

---

[19] Due to licensing restrictions, we cannot disclose the name of this system.

Table 9. Speedup (S) when rating MongoDB, MySQL and SQL-X with M=100K.

Data Store	DBIL	RepairDB	LoadFree	LoadFree+ DBIL	LoadFree+ RepairDB
MongoDB	6.5	3.9	9.1	8.3	6.8
MySQL	11.5	1.9	11.8	11.7	5.1
SQL-X	7.3	3.6	9.2	8.6	6.6

two. This enables BG's rating of MySQL to observe the highest speedups when compared with the naïve technique, see Table 9.

6.3 Speedup with Loading Techniques and DA

This section analyzes the observed speedup with the δ value computed using the DA, highlighting its usefulness to expedite the rating process. We assume the heuristic search technique conducts 11 experiments to rate the data store, comparing the alternative data loading techniques using two different values of δ. First, when δ is set to the SLA duration specified by the experimentalist, i.e., $\delta=\Delta$. Second, when δ is computed using the DA as 16 s. Table 10 shows the latter with different data loading techniques compared with Simple loading technique for four different values of Δ: 3, 10, 100 and 1,000 min. Note that there is a column for Simple, comparing the technique that uses BG to load the database at the beginning of each experiment with the two δ values.

The observed speedup increases as we increase the value of Δ because it causes both f and k to increase, approaching a speedup of 12. With Δ=3 min, Simple observes the lowest speedup as its load time is significant and does not benefit from the use of the DA computing the duration of each experiment to be 11 times faster (δ=16 s instead of 3 min). At the other extreme, LoadFree observes the highest speedup because its database load time is zero and benefits greatly from a δ of 16 s. For the same reason, LoadFree observes approximately the same speedup when Δ approaches 1,000 min.

Table 10. Speedup when rating MongoDB using agile data loading techniques and DA compared to Simple with $\delta=\Delta$.

$\delta = \Delta$ (mins)	Simple	DBIL	RepairDB	LoadFree	LoadFree+DBIL	LoadFree+RepairdDB
3	1.02	7.3	4.2	11	9.8	7.8
10	1.1	7.5	4.3	11.1	9.9	8
100	1.5	8.6	5.6	11.4	10.5	9.02
1,000	4.8	11.03	9.53	11.9	11.6	11.2

7   Related Work

Both the action rating metric and the use of a heuristic search technique to compute this metric for a data store were originally described in [5]. Both were tightly coupled with BG. Subsequently, we realized that both concepts are orthogonal to an application specific benchmark and its abstractions. To illustrate, all concepts described in this paper are applicable to YCSB [13], YCSB++ [24], LinkBench [33] and benchmarks for graph databases [3,18] among the others.

The agile data loading techniques were originally described in [7]. They address the challenge of loading a benchmark database that is a recognized topic by practitioners dating back to Wisconsin Benchmark [8,14] and 007 [10,28,29], and by YCSB [13] and YCSB++ [24] more recently. YCSB++ [24] describes a bulk loading technique that utilizes the high throughput tool of a data store to directly process its generated data and store it in an on-disk format native to the data store. This is similar to DBIL. DBIL is different in two ways. First, DBIL does not require a data store to provide such a tool. Instead, it assumes the data store provides a tool that creates the disk image of the benchmark database once its loaded onto the data store for the very first time. This image is used in subsequent experiments. Second, DBIL accommodates complex schemas similar to BG's schema. Both RepairDB and LoadFree are novel and apply to data stores that do not support either the high throughput tool of YCSB++ or the disk image generation tool of DBIL. They may be adapted and applied to other benchmarking frameworks that rate a data store similar to BG.

To the best of our knowledge, the delta analyzer to compute the duration of each experiment is novel and not described elsewhere in the literature. This component is different than a warmup phase and its duration. Assuming the presence of a warmup phase, it considers how long each experiment must run in order to obtain meaningful numbers.

8   Conclusion

Experimentalists require fast and accurate frameworks to quantify the performance tradeoffs associated with alternative design decisions that implement novel data stores. Response time and throughput are two key metrics. Action rating is a single number that monetizes these two metrics into one. It quantifies the highest observed throughput while a fixed percentage of issued requests observe a response time faster than a pre-specified threshold. When comparing two data stores with one another, the one with the highest action rating is superior. The same holds true when comparing two different algorithms. The primary contribution of this study is a framework to compute action rating. To speedup the rating process, we described the following optimizations: (1) reduce the number of conducted experiments using Approximate instead of Golden with the understanding that its computed SoAR may not be as accurate as Golden, (2) reduce the duration of each conducted experiment using the Delta Analyzer, (3) minimize the time required to create the database at the beginning of each

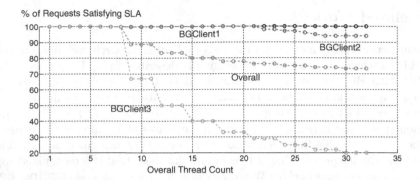

Fig. 11. SoAR rating with three BGClients where one node is slower than the other two.

experiment for those workloads that require it. With these improvements and a fixed amount of time, an experimentalist may rate many more design alternatives to gain insights.

9 Future Work

With the agile data loading techniques, one may expedite the rating process by using a RAID disk subsystem, a RAM disk or even a flash memory. Our immediate future work is to analyze these techniques and to quantify their tradeoffs.

We are extending this study by doing a switch from a closed emulation model to an open one which requires the following changes. First, each BGClient is extended with a main thread that generates requests based on a pre-specified arrival rate using a distribution such as Poisson. This thread generates requests for a collection of threads that issue requests and measure the response time. With Poisson, it is acceptable for a coordinator to require N BGClients to generate requests using an implementation of Poisson with input $\frac{\lambda}{N}$. Now, the heuristic changes the value of λ instead of the number of threads T. The Delta Analyzer uses this same infrastructure to compute the duration of each experiment. The three agile data loading techniques remain unchanged.

Moreover we intend to extend the rating framework by detecting when a component of the framework is a bottleneck. In this case, the obtained action ratings are invalid and the rating process must be repeated with additional computing and networking resources. For example, with the implementation of Sect. 5, there might be an insufficient number of BGClients [5] or one of the BGClients might become the bottleneck because its hardware is slower than that used by other BGClients. For example, Fig. 11 shows the SoAR rating with 3 BGClients where one is unable to produce requests at the rate specified by the BGCoord. This misleads BGCoord and its heuristic to compute 8 instead of 16 as the SoAR rating of the data store. A preventive technique may compute the required number of BGClients given a workload and an approximate processing capability of a data store. Other possibilities include detecting the slower BGClient and either not using it or requiring it to produce a load that does not exhaust its processing capability. These extensions enhance the accuracy of the framework when rating a data store.

References

1. Amdahl, G.: Validity of the Single Processor Approach to Achieving Large Scale Computing Capabilities. In: AFIPS Spring Joint Computer Conference, 30, 483–485 (1967)
2. Amsden, Z., Bronson, N., Cabrera III, G., Chakka, P., Dimov, P., Ding, H., Ferris, J., Giardullo, A., Hoon, J., Kulkarni, S., Lawrence, N., Marchukov, M., Petrov, D., Puzar, L., Venkataramani, V.: TAO: how Facebook serves the social graph. In: SIGMOD Conference (2012)
3. Angles, R., Prat-Pérez, A., Dominguez-Sal, D., Larriba-Pey, J.: Benchmarking database systems for social network applications. In: First International Workshop on Graph Data Management Experiences and Systems, GRADES 2013 (2013)
4. Barahmand, S.: Benchmarking Interactive Social Networking Actions, Doctorate thesis, Computer Science Department, USC (2014)
5. Barahmand, S., Ghandeharizadeh, S.: BG: a benchmark to evaluate interactive social networking actions. In: Proceedings of 2013 CIDR, January 2013
6. Barahmand, S., Ghandeharizadeh, S., Yap, J.: A comparison of two physical data designs for interactive social networking actions. In: Proceedings of the 22nd ACM International Conference on Information and Knowledge Management, CIKM 2013 (2013)
7. Barahmand, S., Ghandeharizadeh, S.: Expedited rating of data stores using agile data loading techniques. In: Proceedings of the 22nd ACM International Conference on Information and Knowledge Management, CIKM 2013, pp. 1637–1642 (2013)
8. Bitton, D., Turbyfill, C., Dewitt, D.J., Systems, B.D.: A systematic approach. In: VLDB, pp. 8–19 (1983)
9. Blasgen, M.W., Gray, J., Mitoma, M.F., Price, T.G.: The convoy phenomenon. Operating Syst. Rev. **13**(2), 20–25 (1979)
10. Carey, M.J., DeWitt, D.J., Naughton, J.F.: The OO7 benchmark. In: SIGMOD Conference, pp. 12–21 (1993)
11. Cattell, R.: Scalable SQL and NoSQL data stores. SIGMOD Rec. **39**, 12–27 (2011)
12. Chang, F., Dean, J., Ghemawat, S., Hsieh, W.C., Wallach, D.A., Burrows, M., Chandra, T., Fikes, A., Gruber, R.E.: Bigtable: a distributed storage system for structured data. ACM Trans. Comput. Syst. **26**(2), 4 (2008)
13. Cooper, B.F., Silberstein, A., Tam, E., Ramakrishnan, R., Sears, R.: Benchmarking cloud serving systems with YCSB. In: Cloud Computing, pp. 143–154 (2010)
14. DeWitt, D., Ghandeharizadeh, S., Schneider, D., Bricker, A., Hsiao, H., Rasmussen, R.: The gamma database machine project. IEEE Trans. Knowl. Data Eng. **1**(2), 44–62 (1990)
15. Ghandeharizadeh, S., Yap, J.: Gumball a race condition prevention technique for cache augmented SQL database management systems. In: Second ACM SIGMOD Workshop on Databases and Social Networks, pp. 1–6 (2012)
16. Gray, J., Reuter, A., Processing, T.: Concepts and Techniques, pp. 677–680. Morgan Kaufmann, Burlington (1993)

17. Gupta, P., Zeldovich, N., Madden, S.: A trigger-based middleware cache for ORMs. In: Kon, F., Kermarrec, A.-M. (eds.) Middleware 2011. LNCS, vol. 7049, pp. 329–349. Springer, Heidelberg (2011)
18. Holzschuher, F., Peinl, R.: Performance of graph query languages: comparison of cypher, gremlin and native access in Neo4J. In: Proceedings of the Joint EDBT/ICDT 2013 Workshops, EDBT 2013, pp. 195–204 (2013)
19. Johnson, R., Pandis, I., Hardavellas, N., Ailamaki, A., Falsafi, B.: Shore-MT: a scalable storage manager for the multicore era. In: EDBT, pp. 24–35 (2009)
20. Lomet, D., Li, F.: Improving transaction-time DBMS performance and functionality. In: IEEE 25th International Conference on Data Engineering, pp. 581–591 (2009)
21. Lomet, D., Vagena, Z., Barga, R.: Recovery from "bad" user transactions. In: SIGMOD, pp. 337–346 (2006)
22. MongoDB. Sharded Cluster Administration (2011). http://docs.mongodb.org/manual/administration/sharded-clusters/
23. MongoDB. Using Filesystem Snapshots to Backup and Restore MongoDB Databases (2011). http://docs.mongodb.org/manual/tutorial/backup-databases-with-filesystem-snapshots/
24. Patil, S., Polte, M., Ren, K., Tantisiriroj, W., Xiao, L., López, J., Gibson, G., Fuchs, A., Rinaldi, B.: YCSB++: benchmarking and performance debugging advanced features inscalable table stores. In: Cloud Computing, New York, NY, USA. ACM (2011)
25. Shute, J., Vingralek, R., Samwel, B., Handy, B., Whipkey, Ch., Rollins, E., Oancea, M., Littlefield, K., Menestrina, D., Ellner, S., Cieslewicz, J., Rae, I., Stancescu, T., Apte, H.: F1: a distributed SQL database that scales. Proc. VLDB Endow. **6**(11), 1068–1079 (2013)
26. Stonebraker, M.: Errors in Database Systems, Eventual Consistency, and the CAP Theorem. Communications of the ACM, BLOG@ACM, April 2010
27. Vogels, W.: Eventually Consistent. Commun. ACM **52**(1), 40–45 (2009)
28. Wiener, J., Naughton, J.: Bulk loading into an OODB: a performance study. In: VLDB (1994)
29. Wiener, J., Naughton, J., OODB bulk loading revisited: the partitioned-list approach. In: VLDB, pp. 30–41 (1995)
30. Jung, H., Han, H., Fekete, A.D., Heiser, G., Yeom, H.Y.: A scalable lock manager for multicores. In: Proceedings of the ACM SIGMOD International Conference on Management of Data, SIGMOD 2013 (2013)
31. Kiefer, J.: Sequential minimax search for a maximum. Proc. Am. Math. Soc. **4**(3), 502–506 (1953)
32. Golden Section Search (2015). http://en.wikipedia.org/wiki/golden_section_search
33. Armstrong, T., Ponnekanti, V., Borthakur, D., Callaghan, M.: LinkBench: a database benchmark based on the facebook social graph. In: SIGMOD, pp. 1185–1196 (2013)

A Theoretical and Experimental Comparison of Filter-Based Equijoins in MapReduce

Thuong-Cang Phan[1]([✉]), Laurent d'Orazio[1], and Philippe Rigaux[2]

[1] Blaise Pascal University, CNRS-UMR 6158-LIMOS, Clermont-Ferrand, France
{ThuongCang.Phan,Laurent.Dorazio}@isima.fr
[2] CNAM, CEDRIC, Paris, France
Philippe.Rigaux@cnam.fr

Abstract. MapReduce has become an increasingly popular framework for large-scale data processing. However, complex operations such as *joins* are quite expensive and require sophisticated techniques. In this paper, we review state-of-the-art strategies for joining several relations in a MapReduce environment and study their extension with *filter-based approaches*. The general objective of filters is to eliminate non-matching data as early as possible in order to reduce the I/O, communication and CPU costs. We examine the impact of systematically adding filters as early as possible in MapReduce join algorithms, both analytically with cost models and practically with evaluations. The study covers binary joins, multi-way joins and recursive joins, and addresses the case of large inputs that gives rise to the most intricate challenges.

Keywords: Big data · Cloud computing · Big data analysis · MapReduce · Equijoin · Bloom filter · Intersection Bloom filter

1   Introduction

Since the advent of applications that propose Web-based services to a world-wide population of connected people, the information technology community has been confronted to unprecedented amount of data, either resulting from an attempt to organize an access to the Web information space (search engines), or directly generated by this massive amount of users (e.g., social networks). Companies like Google and Facebook, representative of those two distinct trends, have developed for their own needs large-scale data processing platforms. These platforms combine an infrastructure based on millions of servers, data repositories where the least collection size is measured in Petabytes, and finally data processing software products that massively exploit distributed computing and batch processing to scale at the required level of magnitude. Although the Web is a primary source of information production, Big Data issues can now be generalized to other areas that continuously collect data and attempt to make sense of it. Sensors incorporated in electronic devices, satellite images, web server logs, bioinformatics, are considered as gold mines of information that just wait for the processing power to be available, reliable, and apt at evaluating complex analytic algorithms.

© Springer-Verlag Berlin Heidelberg 2016
A. Hameurlain et al. (Eds.): TLDKS XXV, LNCS 9620, pp. 33–70, 2016.
DOI: 10.1007/978-3-662-49534-6_2

The MapReduce programming model [13] has become a standard for processing and analyzing large datasets in a massively parallel manner. Its success comes from both its simplicity and nice properties in terms of fault tolerance, a necessary feature when hundreds or even thousands of commodity machines are involved in a job that may extend over days or weeks. However, the MapReduce programming model suffers from severe limitations when it comes to implement algorithms that require data access patterns beyond simple scan/grouping operation. In particular, it is *a priori* not suited for operations with multiple inputs.

One of the most representative such operations are *joins*. A join combines related tuples from datasets on different column schemes and thus raises at a generic level the problem of combining several data sources with a programming framework initially designed for scanning, processing and grouping a single input. Join is a basic building block used in many sophisticated data mining algorithms, and its optimization is essential to ensure efficient data processing at scale.

In the present paper we provide a systematic study of joins with *filters* for early removal of non-participating tuples from the input datasets. As known for a long time in the classical context of relational databases, early elimination of useless data is a quite effective technique to reduce the IO, CPU and communication costs of data processing algorithms. The approach can be transposed in distributed systems in general, and to MapReduce frameworks in particular.

We focus on equijoins, and examine state-of-the-art algorithms for two-way joins, multi-way joins and recursive joins. We compare, analytically and experimentally, the benefit that can be expected by introducing filters as early as possible in the data processing workflow. Our result put the research contributions in this field in a coherent setting and clarifies the stakes of combining several inputs with MapReduce.

The rest of the paper is organized as follows. Section 2 summarizes the background of the basic join operation, recalls the essentials of the MapReduce framework and intersection filters, and positions our paper with respect to related work. Section 3 presents filter-based equijoins in MapReduce. We examine two-way joins, multi-way joins, and recursive joins. Section 4 analyzes the algorithms and introduces cost models. The evaluation environment and the results are reported in Sect. 5. Finally, Sect. 6 concludes and discusses future work.

Table 1 provides a quick reference to the algorithms abbreviations used throughout the text.

2 Background and Related Work

2.1 Join Operation

A join combines tuples from one or several relations according to some join condition[1]. A tuple that participates to the result (and therefore satisfies the join condition) is called a *matching tuple* in the following. Non-matching tuples can simply be ignored from the join processing workflow, a fact that calls for their early elimination. We distinguish the following types of joins:

---
[1] Our study only considers conditions is based an equality operator (=), or *equijoins*.

Table 1. List of abbreviations

Abbreviation	Algorithm
IFBJ	Intersection filter-based join
BJ	Bloom join
RSJ	Reduce-side join
3WJ	Three-way join proposed by Afrati and Ullman [3]
CJ-IFBJ	Chain join using an intersection filter-based join cascade
CJ-BJ	Chain join using a Bloom join (BJ) cascade
CJ-RSJ	Chain join using a reduce-side join (RSJ) cascade
OCJ-2WJ	Optimized chain join using a two-way join cascade
OCJ-3WJ	Optimized chain join using a three-way join (3WJ) cascade
REJ-SHAW	Recursive join using Shaw's approach
REJ-FB	Recursive join using a filter-based approach

- **Two-way join.** Given two datasets R and L, a two-way join denotes the pairs of tuples $r \in R$ and $l \in L$, such that $r.k_1 = l.k_2$ where k_1 and k_2 are join columns in R and L, respectively. The standard notation is:

$$R \bowtie_{k_1 = k_2} L$$

Notation: In order to simplify notations, we will often assume that join keys are known from the context, and will use the abbreviated form $R \bowtie L$.
- **Multi-way join** [35]. Given n datasets $R_1, R_2, \ldots, R_n$, we define a multi-way join as a pairwise combination of two-way joins:

$$R_1 \bowtie R_2 \bowtie R_3 \bowtie \ldots \bowtie R_n$$

Considering only pairwise combination is a restriction: this subclass is sometimes called a *chain join* in the literature.
- **Recursive join** [17,29]. Given a relation $K(x,y)$ encoding a graph, a recursive join computes the transitive closure of K. It requires an initialization, and an iteration (until a fixpoint occurs):

$$\begin{cases} \text{(Initialization)} \ A(x,y) = K(x,y) \\ \text{(Iteration)} \quad A(x,y) = A(x,z) \bowtie K(z,y) \end{cases}$$

We use the following running example: a *user* dataset $R(uid, uname, location)$, a *log* dataset $L(uid, event, logtime)$ and an *acquaintance* dataset $K(uid1, uid2)$. These datasets illustrate the following searches.

- Q_1 - Two-way join. Find the names and events of all users who logged an event before 19/06/2015.

$$A_1(uname, event) = \pi_{uname, event}(R \bowtie \sigma_{logtime < 19/06/2015}(L))$$

- Q_2 - Multi-way join. Find the log events of all users known by Cang

$$A_2(uid, event, logtime) =$$

$$\pi_L(\sigma_{uname='Cang'}(R) \bowtie_{uid=uid1} K \bowtie_{uid2=uid} L)$$

- Q_3 - Recursive join. List the ids of all connected to Philippe.

$$\begin{cases} \text{(Initialization)} & A_3(id) = \pi_{uid}(\sigma_{uname='Philippe'}(R)) \\ \text{(Iteration)} & A_3(id) = \pi_{uid2}(K \bowtie_{uid1=id} A_3) \end{cases}$$

2.2   MapReduce

MapReduce [13] is a parallel and distributed programming model apt at running on computer clusters that scale to thousands of nodes in a fault-tolerant manner. MapReduce usage has become widespread since Google first introduced it in 2004. It allows users to concentrate only on designing their data operations regardless of the distributed aspects of the execution.

A MapReduce job consists of two distinct phases, namely, the *map phase* and the *reduce phase*. Each phase executes a user-defined function on a key-value pair. The user-defined map function **(M)** takes an input pair (k_1, v_1) and outputs a list of intermediate key/value pairs $\langle (k_2, v_2) \rangle$.

$$(k_1, v_1) \xrightarrow{map} list(k2, v2)$$

The intermediate values associated with the same key k_2 are grouped by the framework and then passed to the reduce function which aggregates the values.

$$(k_2, list(v_2)) \xrightarrow{reduce} v_3$$

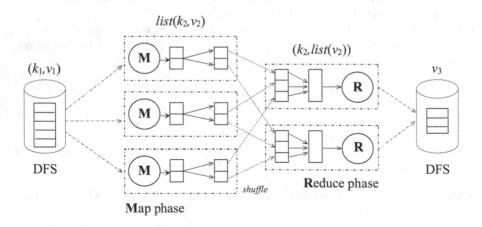

Fig. 1. MapReduce distributed execution

As illustrated by Fig. 1, a typical MapReduce job is executed across multiple nodes. During the map phase, each map task reads a subset (called "*split*") of one input dataset, and applies the map function for each key/value pair of the split. The framework takes care of grouping intermediate data and sends them to the reducer nodes, a communication-intensive process called *shuffling*. Each reduce task collects the intermediate key/value pairs from all the map tasks, sorts/merges the data with the same key, and calls the reduce function to generate the final results.

MapReduce is designed to process a single dataset. Combining several inputs with a MapReduce framework is intricate. The problem has mostly been studied for joins.

2.3 Bloom Filters

A *Bloom filter* (BF) [9] is a space-efficient randomized data structure used for testing membership in a set with a small rate of false positives.

A variant of a Bloom filter is *Intersection Bloom filter* [30], denoted $IBF(S_1, S_2)$, is a probabilistic data structure designed to represent the intersection of sets S_1 and S_2, and check membership in the intersection set. To achieve this, it computes the intersection of the Bloom filters $BF(S_1)$ and $BF(S_2)$. In join processing, matching a join key v against the intersection filter allows to decide (up to the false positive probability) whether it belongs to the shared join keys. The false positive probability of the intersection filter is estimated as f_I representing one of the probabilities of different approaches to the filter [30].

Extended Intersection filter [30] (*EIF*) is developed from the intersection Bloom filter. The *EIF* is a filter built on join key columns $k_1, k_2, \ldots, k_m$ of datasets $R_1, R_2, \ldots, R_m$. It consists of Bloom filters hashed on the key columns, $BF_1(R_1.k_1), BF_2(R_2.k_2 \cap R_3.k_2), \ldots, BF_m(R_m.k_m)$. The membership test takes a tuple $t(k_1, k_2, \ldots, k_m, \ldots, k_n)$ and returns a "*yes*" or "*no*" answer indicating whether t is/is not in the filter. If one of the join keys of the tuple t, $t(k_i)_{i=1\ldots m}$, is not a member of the component filter BF_i of the *EIF*, the output is "*no*" answer. Otherwise, the output is "*yes*" answer. Figure 2 depicts its structure.

For example, consider the three-way join $R(uname, uid) \bowtie K(uid1, uid2) \bowtie L(uid, event)$. K can be filtered by an *EIF* composed of $BF_1(R.uid \cap K.uid1)$ and $BF_2(K.uid2 \cap L.uid)$, i.e., $IBF_1(R.uid, K.uid1)$ and $IBF_2(K.uid2, L.uid)$.

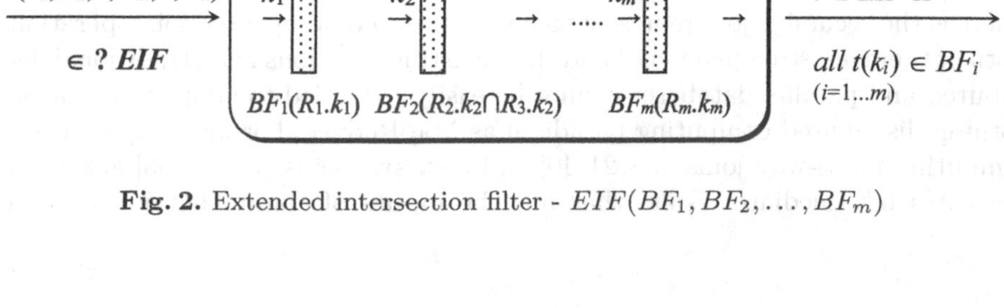

Fig. 2. Extended intersection filter - $EIF(BF_1, BF_2, \ldots, BF_m)$

Each tuple $t(k_1, k_2) \in K$ is checked against the two filters. If k_1 and k_2 are in IBF_1 and IBF_2, respectively, t is accepted, else it is eliminated.

2.4 Joins with MapReduce

Join processing in MapReduce has become a hot research topic in recent years [2,3,8,11,16,22,30]. Many studies have been carried out to evaluate join queries and analyze large datasets in a MapReduce environment. Although joins in MapReduce can be implemented in many ways, the relative performance of the various algorithms depends on certain assumptions such as the size of inputs, data constraints, and joining rates. Map-side joins [8,20,37] would be better to perform the entire joining operation in the map phase since it may save the shuffle and reduce phases. But this solution is limited in running extra MapReduce jobs to repartition the data sources to be usable. Meanwhile, Reduce-side joins [8,20,25,37] are more flexible and general to process a join operation as a standard MapReduce job without any constraints, but they are quite inefficient solutions. Joining does not take place until the reduce phase. In addition, the shuffle phase is really expensive since it needs to shuffle all data, sort and merge.

Observing Reduce-side joins shows that many intermediate pairs generated in the map phase may not actually participate in the joining process due to no matching with any pairs in another input dataset. Consequently, it would be much more efficient if we eliminate the non-matching data right in the map phase. This problem can be solved by Semi-join [8]. It uses a distributed cache to disseminate a hashmap of one of input datasets across all the mappers, then dropping tuples whose join key not in the hashmap. The main obstacle in this way resides at the hashmap because the hashmap may not fit in memory and its replication across all the mappers may be inefficient. In this situation, therefore, Bloom join [19,22,23,39,40] is a worthy replacement for Semi-join because it benefits from a Bloom filter [9] to do existence tests in less memory than a full list of keys from the hashmap. Another restriction on these solutions is derived from their filtering efficiency, even for recent research efforts [3,22,40]. There remain a lot of non-matching data after filtering because the solutions can only filter on one of input datasets instead of both. Thus, Intersection filter-based join [30] may become a better solution to address this problem by eliminating non-matching data from both input datasets. However, it is necessary to have a complete evaluation of the solutions that indicates their benefits and limitations.

In addition to the above two-way joins, the researchers are also confronted big challenges that come from multi-way joins and recursive joins in MapReduce. The multi-way join extends the two-way join by handling multiple input datasets, whereas the recursive join represents a computation of a repeated join operation. Both of them are still open issues and their existing solutions from traditional distributed and parallel databases cannot be easily extended to adapt to a shared-nothing distributed computing paradigm as MapReduce. For latest approaches, computing multi-way joins [3,8,21,40] and recursive joins [1,2,12,33] also often generates intermediate results that may be inputs of component joins of the joins. These intermediate results contain a lot of non-matching data that considerably increases total overheads for I/O, CPU, sort, merge, and especially communication. We need to figure out optimized solutions that can prevent the non-matching data involved in the intermediate results. Besides, minimizing the intermediate data amount sent to the reducers should be calculated appropriately.

The purpose of the present paper is to provide a consistent review of filter-based join processing techniques in a MapReduce environment. It not only (a) covers the recently various techniques for computing two-way joins, multi-way joins and recursive joins, but also (b) qualifies these techniques with cost models and (c) evaluates them with experimental studies to both validate the proposed cost model and investigate their practical behavior. Overall, our goal is to provide a clear, robust and comparative assessment of join processing solutions to guide the choice of practitioners confronted to the need to perform join at scale in a specific context. By founding this assessment on both an analytic and empirical study, we hope to provide a material that puts the research contributions in this field in a coherent setting and clarifies the stakes of combining several inputs with MapReduce.

For the sake of consistency, we focus on join algorithms that share some common features. First, we only consider equijoins. Second, we investigate algorithms that exploit filters to reduce the network communication. Filtering is a strategy that can be combined with all kind of approaches, and turns out to be (almost) always beneficial in a context where I/O and network exchanges constitute the major bottleneck. Third, our work complements a few other surveys recently published [14, 20, 24, 31, 32, 40] which, on the one hand, explore a larger scope (e.g., non-equi joins [14, 27, 38, 41]), but on the other hand do not propose an in-depth coverage as we do, and a comparison methods uniformly applied to the range of proposals published so far.

3 Extending Equijoins with Filters in MapReduce

The most straightforward way to join datasets with MapReduce is the Reduce-side join algorithm [8, 20, 25, 37], denoted RSJ. It groups tuples from both datasets on their respective join key value during the map phase, and merges/joins them during the reduce phase. Tuples are processed regardless of their actual contribution to the final result, and thus the join algorithm has to pay an overhead for processing and shipping useless data.

Consider for instance the Facebook user dataset R containing more than 1.23 billion users [15]. We would like to obtain users' activities in a certain period of time (e.g., one hour) by joining R and the log dataset L. Since L, *over this period*, contains the activities of only a few million unique users, most of the users in R are not represented, and RSJ spends useless resources to access, process and transfer over the network the non-matching tuples of R.

Several *filter-based* extensions have been proposed to tackle the problem. Their common idea is to filter out the non-matching tuples from the input datasets during the map phase. A *filter* in this context is a compact data structure that supports fast membership tests. Filter-based joins require two stages:

- *Stage1* (pre-processing). A filter F is built on a join key value set of one input dataset. For the intersection filter, F represents the intersection of the key value sets. A membership test for some key value k on F tells whether k participates or not to the join result.
- *Stage 2* (join). F is distributed to all the computing nodes, and used to eliminate non-matching tuples during the map phase. The join then proceeds as explained above.

A filter is a compact representation of a set. It accepts a rate of false positives (i.e., positive answer for non-matching tuples in our case) but no false negatives. Filtering avoids the communication overhead of shipping tuples from the mappers to the reducers, and the storage and CPU overhead of processing such tuples during the reduce phase. The join strategy remains unchanged, and exploits the MapReduce paradigm: the input datasets are partitioned and grouped during the map phase, in order to solve locally the problem during the reduce phase. Filtering presents some advantages and disadvantages:

- *Advantages*: the strategy does not impose any restrictions on input datasets, nor modifications to the MapReduce framework. Besides, it removes non-matching data to reduce the communication overhead.
- *Disadvantages*: building the filters represents a significant cost, since it requires scanning the input, and transferring the filters.

In the rest of this section, we examine in detail the application of filter-based techniques to the following join variants: two-way joins, multi-way joins and recursive joins. For each variant, we present the state-of-the-art algorithms, along with a discussion on their expected advantages/disadvantages.

3.1 Two-Way Joins

A two-way join $R_1 \bowtie R_2$ involves two relations R_1 and R_2. In the following r_1 (resp. r_2) denote a tuple from R_1 (resp. R_2) and k refers to the join key attribute. We use simplified notations when allowed by the context.

Bloom Joins. *Bloom join* (BJ) [19,22,39] is a specific type of the filter-based join strategy in which the well-known Bloom filter [9] is used. BJ is implemented by two MapReduce jobs as follows:

- *Job 1* (preprocessing) is a job with only one reducer. The mappers scan splits of the input R_2, extract the join key value from each tuple, and produces local Bloom filters. Then, the mappers emit the local filters to the reducer that merges them into a global filter $BF(R_2)$ using the bit-wise **OR**.

- *Job 2* (processing) filters out non-matching tuples in R_1 and joins the filtered result R_1' with R_2. It relies on a distributed cache to store $BF(R_2)$. The mappers scan splits of R_1 and R_2, and eliminate the tuples of R_1 whose keys are not in $BF(R_2)$. Tuples from R_2 are not filtered.
  Each tuple is then ticked with a tag that indicates its dataset name. For our example, mappers emit tagged tuples with composite keys of the form $((r_1.k, 'R_1'), r_1)$ or $((r_2.k, 'R_2'), r_2)$. The reducers receive tagged tuples grouped on the k value (this requires a small change of the partitioning function). For each group, the reduce function constructs all the pairs (r_1, r_2) to complete the join.

Note that it requires to override the default grouping function in order to ensure that grouping the tagged tuples takes into consideration only the join key part and ignores the tag part. The tag is used for secondary sort which ensures that, for a given key value, all tuples from R_1 are processed before those of R_2. This allows to apply a standard in-memory hash join.

Discussion. BJ benefits from the compacity of the Bloom filter to reduce the amount of data transferred over the network. The size of the filter can be fixed regardless of the number of join keys. However, given a fixed filter size, the probability f of false positives increases with the number of join keys.

A major concern with the filtering approach in general is the need to run a pre-processing job for building the filter. Besides, broadcasting the filter becomes inefficient if its size is large. Finally, it is worth noting that the BJ is asymmetric: non-matching tuples of R_2 have not been filtered, hence the problem is half-solved.

The authors of [22] have proposed an improvement of BJ that avoids the pre-processing job, but requires two internal modifications of the framework. We do not consider in the present study such extensions that necessitate a non-standard environment.

Intersection Filter-Based Joins. We now describe an improvement of the above approach, the *Intersection filter-based join* [30], denoted IFBJ. It relies on the fact that only tuples whose join keys belong to the set of *shared* join keys do participate to the result.

The implementation of IFBJ is done with the following jobs:

- *Job 1* (pre-processing) is a job with only one reducer. The mappers scan splits of R_1 and R_2, extract the join key value for each tuple, and insert them in the local Bloom filters regardless of duplicate keys.
  The mappers then emit the local filters to the reducer which merges them in two global filters $BF(R_1)$ and $BF(R_2)$ using the bit-wise OR. Based on one of three approaches to building the intersection filter [30], the reducer computes the intersection filter $IBF(R_1, R_2)$ from the global filters.
- *Job 2* (join) uses a distributed cache to provide IBF to all the compute nodes. The mappers scan splits of R_1 and R_2, extract the join key for each tuple and

match it against the intersection filter. If the key v belongs to the intersection filter, the tuple is emitted as a pair $((v, tag), tuple)$. The join evaluation in the reduce phase is similar to the Bloom join algorithm.

IFBJ benefits from the standard features of Bloom filters: its small size, its independence from the number of the keys and key duplication, and fast membership test. Join based on the intersection filter is expected to be more efficient than the Bloom join because of its ability to filter out non-matching tuples from both two input datasets. An interesting characteristic of the intersection filter is that if $IBF(R_1, R_2)$ has all bits set to zero, then the sets $R_1.k$ and $R_2.k$ are disjoint and the join evaluation stops without doing anything. However, the algorithm has to pay the additional cost of a MapReduce job for building the intersection filter and requires scanning the two input datasets twice.

3.2   Multi-way Joins

We can extend the above approach to the computation of multi-way joins with an extended intersection filter (EIF) in the following.

Three-Way Joins. We begin our study of multi-way joins by considering the special case of a three-way join $R_1 \bowtie R_2 \bowtie R_3$. For the sake of concreteness, we will discuss the following query on our example relations.

$$R \bowtie_{uid=uid1} K \bowtie_{uid2=uid} L$$

There are several possible pairwise combinations to compute this three-way join.

$$R \bowtie_{uid=uid1} K \bowtie_{uid2=uid} L = (R \bowtie_{uid=uid1} K) \bowtie_{uid2=uid} L$$
$$= R \bowtie_{uid=uid1} (K \bowtie_{uid2=uid} L)$$

We can evaluate three-way joins as a sequence of 2 two-way joins, using two successive jobs. An alternative is to join the three datasets together with a

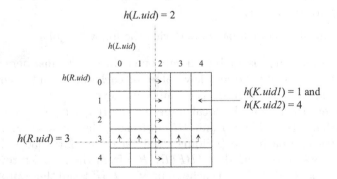

Fig. 3. Distributing tuples of R, K, and L among $r = n^2$ reducers

single job, as recently proposed by Afrati and Ullman [3]. It relies on the idea of a *matrix of reducers* as shown in Fig. 3.

The number of reducers must be the square of some integer n ($r = n^2$) and reducers are mapped (virtually) to a matrix $n \times n$. Each reducer is mapped to a cell (i, j), and identified by $i * n + j$. With $n = 5$, cell $(3, 2)$ is for instance associated with the reducer 17.

The mappers assign tuples of R, K, and L to the reducer matrix as follows. Let h be a hash function with range $[0, n - 1]$. Each tuple of K is sent to a single reducer, the one in cell $(h(K.uid1), h(K.uid2))$. Tuples from R and L are sent to all the reducers of, resp. a whole row or column in the matrix. Each tuple $r(uid, uname)$ is sent to all the reducers of the *row $h(uid)$*. Each tuple $l(uid, event)$ is sent to all the reducers of the *column $h(uid)$*.

We can give a perspective: assume three tuples $R('Laurent', u_1)$, $K(u_1, u_2)$, and $L(u_2, 'login')$. They will all be sent to the reducer $h(u_1) * n + h(u_2)$ and the joined tuple will therefore be produced.

Let us assume, for simplicity, that $|R|=|K|=|L|$. The total communication cost for the Afrati's three-way join (denoted 3WJ in the following) is $O(|R|.\sqrt{r})$, whereas the total communication cost for the cascade of 2 two-way joins without filters is $O(|R|^2.\alpha)$, where α is the probability for two tuples from different datasets to match on the join key (Sect. 4.2 for more details). It follows that 3WJ is better than the cascade of the two-way joins when $r < (|R|.\alpha)^2$.

A downside of 3WJ is that it generates n duplicates for each tuple of either R or L. This represents a large communication and I/O overhead. This situation can be improved significantly by removing non-matching tuples prior to the reduce phase. We extend 3WJ with intersection filters as shown in Fig. 4.

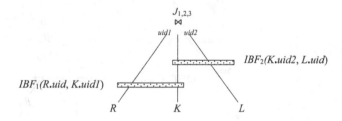

Fig. 4. Three-way join extended with intersection filters

R and L are filtered by $IBF_1(R.uid, K.uid1)$ and $IBF_2(K.uid2, L.uid)$, respectively. K is filtered by an extended intersection filter $EIF(IBF_1, IBF_2)$.

The extension of the three-way join with filters uses two jobs as follows.

- *Job 1* (pre-processing) builds $IBF_1(R.uid, K.uid1)$ and $IBF_2(K.uid2, L.uid)$. Let mp_1, mp_2 and mp_3 be the number of map tasks for R, K and L, respectively. The job consists of $mp_1 + mp_2 + mp_3$ map tasks that build filters and one reduce task that produces two intersection filters. In detail, mp_1 tasks build local filters on $R.uid$; mp_2 tasks build local filters on $K.uid1$ and $K.uid2$; mp_3 tasks build local filters on $L.uid$. Those filters are shipped to the reducer which produces $BF(R.uid)$, $BF(L.uid)$, $BF(K.uid1)$, $BF(K.uid2)$, as well as IBF_1, IBF_2 and $EIF(IBF_1, IBF_2)$. Note that the join result is known to be empty right away if either IBF_1 or IBF_2 is empty.

- *Job 2* (join) filters out non-matching tuples from R, K and L, and carries out the join evaluation. It distributes the intersection filters to all tasktrackers, creates map tasks for R, K and L and r reduce tasks.

    ⋆ *Map phase with filtering:* Each mapper matches any tuple of R or L against the relevant filter IBF_1, IBF_2, or $EIF(IBF_1, IBF_2)$. Tuples that pass the filtering process are then sent to the reducers according the 3WJ policy. This involves tuple replication as shown in Fig. 3.

    ⋆ *Reduce phase:* the reduce function applies a full cross-product of tuples from the different input datasets. Locally, the reducer buffers the tuples of R and L, streams the tuples from K, and performs the cross product.

Chain Joins. We now consider the more general case of multi-way joins, or *chain joins*, a sequence or pair wise joins of the form of $R_1(x_1, x_2) \bowtie R_2(x_2, x_3) \bowtie R_3(x_3, x_4) \bowtie \cdots \bowtie R_n(x_n, x_{n+1})$.

The baseline solution is a cascade of Bloom joins (CJ-BJ). The query plan is a left-deep join tree, and relies on a set of filters $BF_2(R_2.x_2), \ldots, BF_n(R_n.x_n)$ built on the base datasets by a pre-processing job. In this scenario, we can regconize that R_1 and all intermediate results $R_{1,2,\ldots,i}$ are filtered by the filters, whereas the base relations R_i are not, where $i \in [2, n]$.

We propose an improved evaluation that generalizes intersection filters as shown by Fig. 5. In addition to the filters BF on base relations, the extended

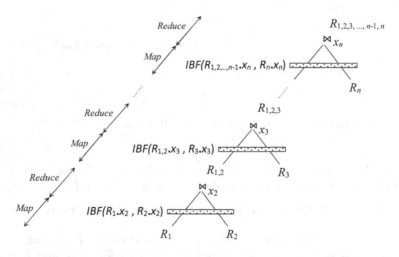

Fig. 5. Implementation of a chain join using a cascade of two-way joins using intersection filters (CJ-IFBJ)

algorithm denoted CJ-IFBJ creates on the fly intersection Bloom Filters on intermediate results, $IBF(R_{1,\cdots,i-1}.x_i, R_i.x_i), i \in [2,n]$ during the reduce phases of intermediate joins

All the input datasets and intermediate join results are filtered by their corresponding intersection filters. For instance, $IBF(R_{1,2}.x_3, R_3.x_3)$ is used to eliminate non-matching tuples in both $R_{1,2}$ and R_3. Intermediate data sent to the reducers with CJ-IFBJ is expected to be much less than in the case of CJ-BJ.

We can even go one step further by noting that intermediate join results still contain non-matching tuples transmitted to the next join. For instance, the join of R_1 and R_2 likely contains tuples that do not match any tuples of R_3 on x_3. We can therefore "push" the filter $BF(R_3.x_3)$, down to the scan of relation R_2. The idea is actually quite reminiscent of the traditional optimization heuristics that pushes selection down the query tree in relational systems.

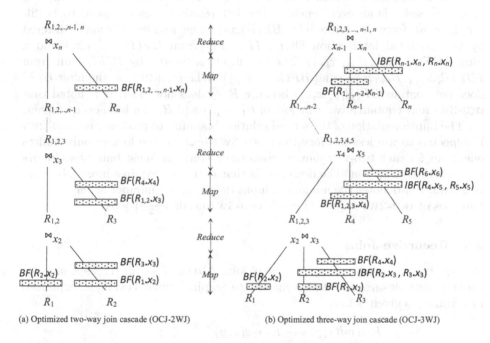

(a) Optimized two-way join cascade (OCJ-2WJ)   (b) Optimized three-way join cascade (OCJ-3WJ)

Fig. 6. Optimization of a chain join using extended intersection filters (OCJ)

Figure 6(a) shows a first optimized solution using a cascade of filter-based two-way joins, denoted OCJ-2WJ. The input datasets $R_2, \ldots, R_n$ are filtered by extended intersection filters EIF. The extended filter EIF_i includes a filter $BF(R_{1,\cdots,i-1}.x_i)$ built from the intermediate join result and a filter $BF(R_{i+1}.x_{i+1})$ from the next input dataset, where $i \in (2,n)$. Specially, EIF_2 contains $BF(R_1.x_2)$ and $BF(R_3.x_3)$, and EIF_n only consists of $BF(R_{1,2,\cdots,n-1}.x_n)$. This solution ensures that intermediate join results only contain (up to false positives) matching data that can be sent to the next join without filtering. This is an important characteristic which avoids to apply additional filters to intermediate join results.

The implementation first uses a pre-processing job to build the Bloom filters $BF(R_i.x_i)$, $i = 2, \ldots, n$, and $BF(R_1.x_2)$. Next, it evaluates the chain join as a sequence of two-way joins. During the evaluation of $R_{1,\ldots,i-1} \bowtie R_i$, the left input need not be filtered, except R_1 filtered by $BF(R_2.x_2)$. The right input is filtered by the EIF_i built from $BF(R_{1,\ldots,i-1}.x_i)$ and $BF(R_{i+1}.x_{i+1})$. The former is generated in the reduce phase of the previous join processing between $R_{1,\ldots,i-2}$ and R_{i-1}. Building the filters from the intermediate join results does not involve any overhead. The iteration stops if one of the two input datasets is null.

Figure 6(b) illustrates a second optimization, where pairwise joins are replaced by filtered three-way joins (3WJ). We denote this further optimized solution as OCJ-3WJ. Consider the three-way join $R_{1,\ldots,i-1} \bowtie R_i \bowtie R_{i+1}, i \in [2, n-1]$ and i is an even number. The left relation does not need to be filtered, apart from R_1 filtered by $BF(R_2.x_2)$. The middle relation is filtered by the extended intersection filter EIF_i built from $BF(R_{1,\ldots,i-1}.x_i)$ and a filter $IBF(R_i.x_{i+1}, R_{i+1}.x_{i+1})$. The last input is filtered by EIF'_i, built from $IBF(R_i.x_{i+1}, R_{i+1}.x_{i+1})$ and $BF(R_{i+2}.x_{i+2})$. When $(i+2) > n$, the filter EIF'_i does not contain $BF(R_{i+2}.x_{i+2})$ because R_{i+2} does not exist. It is noted that OCJ-3WJ may contain a two-way join of $R_{1,\ldots,n-1}$ and R_n if n is an even number.

The implementation of the second solution is similar to the first one. OCJ-2WJ is expected to use less memory than OCJ-3WJ because the former only buffers one input for each two-way join, whereas the second one must buffer two inputs for each three-way join. The downside is that OCJ-2WJ requires more jobs than OCJ-3WJ. If n denotes the number of input datasets, the number of the two-way join jobs of OCJ-2WJ is $(n-1)$, while OCJ-3WJ needs $\lceil \frac{n-1}{2} \rceil$ jobs.

3.3 Recursive Joins

We now turn to another complex type of join. A *recursive join* [17,29] computes the transitive closure of a relation encoding a graph. A typical example, expressed in Datalog, is given below.

$$Friend(x, y) \longleftarrow Know(x, y)$$
$$Friend(x, y) \longleftarrow Friend(x, z) \bowtie Know(z, y)$$

Evaluating a recursive join is tantamount to computing the transitive closure of the graph represented by the relation. This can be done via an iterative process that stops whenever a fixpoint is reached. We examine how the semi-naive algorithm [36] can be evaluated in MapReduce.

Let F and K denote the relations *Friend* and *Know*, respectively. Let $F_i, 0 < i \leq n$ be the temporary value of the relation *Friend* at step 0, with $F_0 = \emptyset$. The incremental relation of $F_i, i > 0$, denoted ΔF_i, is defined as:

$$\Delta F_i = F_i - F_{i-1} = \Pi_{xy}(\Delta F_{i-1} \bowtie_z K) - F_{i-1}$$

The semi-naive algorithm uses this delta relation to avoid redundant computations (Algorithm 1).

Algorithm 1. Semi-Naive evaluation for recursive joins

Input: A graph encoded as a relation K
Output: The transitive closure of K
1 $F = \emptyset, \Delta F_0 = K(x, y), i = 1$
2 **while** $\Delta F_{i-1} \,! = \emptyset$ **do**
3 | $\Delta F_i = \Pi_{xy}(\Delta F_{i-1} \bowtie_z K) - F$
4 | $F = F \cup \Delta F_i$
5 | $i + = 1$
6 **return** F

At each step i, some new facts are inferred and stored in ΔF_i. The loop is repeated until no new fact is inferred ($\Delta F_i = \emptyset$), i.e., the fixpoint is reached. The union of all incremental relations, ($\Delta F_0 \cup \ldots \cup \Delta F_{i-1}$), is the transitive closure of the graph.

Shaw et al. [33] have proposed the following algorithm to implement the semi-naive algorithm in MapReduce (REJ-SHAW). Each iteration evaluates ΔF_i = $\Pi_{xy}(\Delta F_{i-1} \bowtie_z K) - F_{i-1}$ with two jobs, namely, one for join and one for deduplication and difference (*dedup-diff*), as shown on Fig. 7.

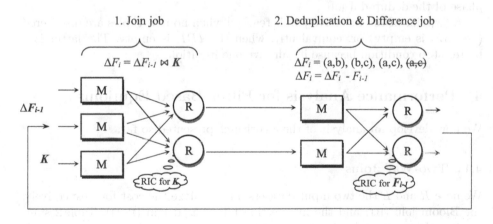

Fig. 7. Semi-naive implementation of recursive joins in MapReduce

The first job computes ($\Delta F_{i-1} \bowtie K$), the second computes the new delta relation ΔF_i. This 2-jobs execution is iterated until ΔF_i is empty. This means that the invariant relation K and the incremental relation F_{i-1} are re-scanned and re-shuffled for every iteration. Shaw et al. have tackled this situation in the HaLoop system [12] by using the Reducer Input Cache (RIC). RIC stores and indexes reducer inputs across all reducers. To avoid re-scanning and re-shuffling the same data with the same mapper on iterations, the solution therefore uses RIC for the datasets K and F_{i-1} in the join job and the dedup-diff job, respectively, as shown on Fig. 7. K is scanned only once, at the first loop. K_i and K_j are splits of K, which are cached at the reducer input caches i and j, resp. Note that caching intermediate results during iterative computations is now integrated in modern distributed engines such as Spark [7] and Flink [5].

The dedup-diff job using RIC is described as follows. Each tuple is stored in the cache as a key/value pair (t, i), where the key is the tuple t discovered by the previous join job and the value is the iteration number i for which that tuple was discovered. The map phase of the difference job hashes the incoming tuples as keys with values indicating the current iteration number. During the reduce phase, for each incoming tuple, the cache is probed to find all instances of the tuples previously discovered across all iterations. Both the incoming and cached data are passed to the user-defined reduce function. A tuple previously discovered is omitted from the output, else it is included in ΔF_i.

When evaluating $(\Delta F_{i-1} \bowtie K)$, Shaw's solution (REJ-SHAW) does not discover and eliminate non-matching tuples from ΔF_{i-1} and K. Our extension, REJ-FB in the following, adds an intersection filter $IBF(\Delta F_{i-1}.z, K.z)$ as proposed in Sect. 2.3. Initially, the filter is simply $BF(K.z)$ generated by a pre-processing job. During the ith iteration ($i \geq 1$), REJ-FB uses $IBF(\Delta F_{i-1}.z, K.z)$ as a filter in the map phase of the join job, and builds $IBF(\Delta F_i.z, K.z)$ in the reduce phase of the dedup-diff job.

A fixpoint of the recursive join is reached when no new tuples are discovered (i.e. ΔF_i is empty) or, equivalently, when the IBF is empty. The latter is a better stop condition because it can save one iteration.

4   Performance Analysis for Filter-Based Equijoins

We now develop an analysis of the algorithms presented so far.

4.1   Two-Way Joins

We note R and L the two input datasets, and analyze the cost for, respectively, the Bloom join (BJ) and the intersection filter-based join (IFBJ). Table 2 summarizes the parameters of our cost model.

Cost Model. We adapt the cost model presented in [26]. We propose the following global formula that captures the cost of a two-way join.

$$C_{2wJoin} = C_{pre} + C_{read} + C_{sort} + C_{tr} + C_{write} \qquad (1)$$

where:

- $C_{read} = c_r \cdot |R| + c_r \cdot |L|$
- $C_{sort} = c_l \cdot |D| \cdot 2 \cdot (\lceil \log_B |D| - \log_B(mp) \rceil + \lceil \log_B(mp) \rceil)$ [26]
- $C_{tr} = c_t \cdot |D|$
- $C_{write} = c_r \cdot |O|$
- $C_{pre} = C' + c_r \cdot m \cdot t$

  ◇ $C' = \begin{cases} C_{read} + (c_l + c_t) \cdot m \cdot mp & \text{, for IFBJ} \\ c_r \cdot |L| + (c_l + c_t) \cdot m \cdot mp_2 & \text{, for BJ} \end{cases}$

  ◇ $C_{pre} = 0$, for approaches without filters. In addition, it is assumed that the filters are the same size m. If m is small, we will not compress the filter files and m is therefore the size of the Bloom filter.

An additional component, C_{pre}, is added to the cost model in [26] to form Eq. (1). $|D|$, the size of the intermediate data, strongly influences the total cost of a join operation, and is essential in particular to decide whether the filter-based variant of the algorithm is worth its cost.

Table 2. Parameters of the cost model for two-way joins

Parameter	Explanation
$|R|$	The size of R
$|L|$	The size of L
$|D|$	The size of the intermediate data
c_l	The cost of reading or writing data locally
c_r	The cost of reading/writing data remotely
c_t	The cost of transferring data from one node to another
$B+1$	The size of the sort buffer in pages
mp_1	The number of map tasks for R
mp_2	The number of map tasks for L
mp	The total number of map tasks, $mp = mp_1 + mp_2$
t	The number of tasktrackers
m	The compressed size of the Bloom filter (bits) m = the size of the Bloom filter $\times$ the file compression ratio
$|O|$	The size of the join processing output
C_{pre}	The total cost to perform the pre-processing job
C_{read}	The total cost to read the data
C_{sort}	The total cost to perform the sorting and copying at the map and reduce nodes
C_{tr}	The total cost to transfer intermediate data among the nodes
C_{write}	The total cost to write the data on DFS

Cost Comparison. In this section, we evaluate $|D|$, for each algorithm mentioned in Sect. 3.1, and provide a cost comparison. Importantly, we identify a threshold that can guide the choice amongst of these algorithms. We add the Reduce-side join (RSJ) to our comparison to highlight the effect of filtering.

We denote as δ_L and δ_R, respectively, the ratio of the joined records of R with L (resp. L with R). The size of intermediate data is:

$$|D| = \begin{cases} \delta_L|R| + f_I(R,L) \cdot (1 - \delta_L)|R| + \delta_R|L| + f_I(R,L) \cdot (1 - \delta_R)|L| & (2) \\ \delta_L|R| + f(L) \quad\quad \cdot (1 - \delta_L)|R| + \quad |L| & (3) \\ |R| \quad\quad\quad\quad\quad\quad\quad\quad\quad\quad + \quad |L| & (4) \end{cases}$$

where:

- Equation (2) holds for IFBJ, denoted D_{IFBJ}
- Equation (3) holds for BJ, denoted D_{BJ}
- Equation (4) holds for RSJ, denoted D_{RSJ}
- $f_I(R,L)$ is the false positive probability of the intersection filter $IBF(R,L)$ [30],
- and $f(L)$ is the false positive probability of the Bloom filter $BF(L)$.

From these equations, we can infer the following.

Theorem 1. *An IFBJ is more efficient than a BJ because it produces less intermediate data. Additionally, the following inequality holds:*

$$D_{\text{IFBJ}} \leq D_{\text{BJ}} \leq D_{\text{RSJ}} \tag{5}$$

where D_{IFBJ}, D_{BJ}, and D_{RSJ} are the sizes of intermediate data of IFBJ, BJ, and RSJ, resp.

Proof. We get $0 < f_I(R,L) < f(L) < 1$ [30]. So we can deduce that:

$$\delta_L \cdot |R| + f_I(R,L) \cdot (1 - \delta_L) \cdot |R| \leq \delta_L \cdot |R| + f(L) \cdot (1 - \delta_L) \cdot |R| \leq |R| \text{ and} \tag{6}$$

$$\delta_R \cdot |L| + f_I(R,L) \cdot (1 - \delta_R)|L| \leq |L| \tag{7}$$

Note that the ratio of the joined records, δ_L or δ_R, could be 1 in the case of a join based on a foreign key.

By combining inequalities (6) and (7) into Eqs. (2), (3) and (4), Theorem 1 is proved. □

From Eqs. (1) and (5), we can evaluate the total cost of the join operation for the different approaches.

Theorem 2. *Once the pre-processing cost C_{pre} is negligible or less than the cost of non-matching data, an IFBJ has the lowest cost. In addition, a comparison of the costs is given by:*

$$C_{\text{IFBJ}} \leq C_{\text{BJ}} \leq C_{\text{RSJ}} \tag{8}$$

where C_{IFBJ}, C_{BJ}, and C_{RSJ} are the total costs of IFBJ, BJ, *and* RSJ, *resp. As a result, the most efficient join approach is typically* IFBJ, *the second one is* BJ, *and the worst one is* RSJ.

The total cost to perform the pre-processing job is given by:

$$C_{pre} = \begin{cases} C_{read} + (c_l + c_t) \cdot m \cdot mp + c_r \cdot m \cdot t & \text{, in case of IFBJ} \\ c_r \cdot |L| + (c_l + c_t) \cdot m \cdot mp_2 + c_r \cdot m \cdot t & \text{, in case of BJ} \\ 0 & \text{, in case of RSJ} \end{cases}$$

Regarding data locality, the MapReduce framework makes its best efforts to run the map task on a node where the input data resides. Although this cannot always be achieved, we can see that the cost of this phase, C_{pre}, is negligible compared to the generation and transfer of non-matching tuples over the network. In general, choosing the filter-based joins relies on the read cost c_r and a threshold of non-matching data shown in Theorem 3.

The filter-based join algorithms will become inefficient when there is a large number of map tasks, and very little non-matching data in the join operation. For large inputs with many map tasks, a tasktracker running multiple map tasks will maintain only two local filters $BF(R)$ and (or) $BF(L)$ thanks to merging the local filters of the tasks. Two hundred map tasks running on a tasktracker, for instance, will produce 200 local filters $BF(R)$. The tasktracker merges all the local filters into one $BF(R)$. Besides, as the number of non-matching tuples decreases, the filters become useless and computing them with an additional job represents a penalty. It hence needs to indicate the dependence of the filter-based joins on the amount of non-matching data through estimating the threshold for this data that determines whether filters should be used.

Let $|D^*|$ be the size of non-matching data, C^*_{sort} be the total cost of sorting and copying it at the map and reduce nodes, and C^*_{tr} be the total cost to transfer it among the nodes. Accordingly, the cost associated with non-matching data is the sum of C^*_{sort} and C^*_{tr}.

Theorem 3. *The filter-based joins become a good choice when:*

$$C_{pre} < C^*_{sort} + C^*_{tr} \tag{9}$$

where:

- $|D^*| = |R| + |L| - |D|$
- $C^*_{tr} = c_t \cdot |D^*|$
- $C^*_{sort} = c_l \cdot |D^*| \cdot 2 \cdot (\lceil \log_B |D^*| \rceil - \log_B(mp)\rceil + \lceil \log_B(mp) \rceil)$ [26]

Based on the size of $|D|$, the threshold depends on δ_L and δ_R (the ratio of the joined records).

In summary, the best choice of the join approaches is IFBJ, the second one is BJ, and the worst one is RSJ (Theorem 2). However, this would become incorrect when the join has small input datasets and a high ratio of matching tuples that is defined by the threshold of the joined records (Theorem 3). In these cases, RSJ would be the best choice and the filter-based joins should not be used because the cost of building and broadcasting filter(s) becomes relatively significant.

4.2 Multi-way Joins

Three-Way Joins. Let R, K and L be three input datasets. The general formula that estimates the total cost of 3WJ is:

$$C_{3wJoin} = C_{pre} + C_{read} + C_{sort} + C_{tr} + C_{write} \tag{10}$$

where:

- $C_{read} = c_r \cdot |R| + c_r \cdot |K| + c_r \cdot |L|$
- $C_{sort} = c_l \cdot |D| \cdot 2 \cdot (\lceil \log_B |D| - \log_B(mp) \rceil + \lceil \log_B(mp) \rceil)$ [26]
- $mp = mp_1 + mp_2 + mp_3$, the total number of map tasks for the three inputs
- $C_{tr} = c_t \cdot |D|$
- $C_{write} = c_r \cdot |O|$
- $C_{pre} = C_{read} + (c_l + c_t) \cdot m \cdot mp + 2 \cdot c_r \cdot m \cdot t$, for 3WJ using filters;
  $C_{pre} = 0$ for 3WJ.

To simplify the computation, we suppose that R, K and L have the same size. A 3WJ increases the communication cost because each tuple of R and L is sent to many different reducers. On the other hand, the two-way join cascade must launch an additional job, then scan and shuffle the intermediate result. We characterize the relative costs of the approaches as follows.

Theorem 4. *A* 3WJ, *$R(A, B) \bowtie K(B, C) \bowtie L(C, D)$, is more efficient than a cascade of 2 two-way joins $(R(A, B) \bowtie K(B, C)) \bowtie L(C, D)$ or $R(A, B) \bowtie (K(B, C) \bowtie L(C, D))$ when $r < (|R|.\alpha)^2$. Additionally, the size of the intermediate data is specified by*

$$|D| = \begin{cases} 2 \cdot |R| \cdot \sqrt{r} & \text{, for 3WJ.} \\ |R|^2 \cdot \alpha & \text{, for a cascade of 2 two-way joins.} \end{cases}$$

where r is the number of reducers, $|R| = |K| = |L|$, and α is the probability of two tuples from different datasets to match on the join key column.

Proof. Similar to the proof of Afrati and Ullman in [3]. First, we consider 3WJ. Two attributes B and C are join key columns. We use hash functions to map values of B to b different buckets, and values of C to c buckets, as long as $b \cdot c = r$. The intermediate data size of the three-way join is

$$|R| \cdot c + |K| + |L| \cdot b \tag{11}$$

We must find optimal values for b and c to minimize the above expression subject to the constraint that $b \cdot c = r$, b and c being positive integers. In this case, the Lagrangian multiplier method is used to present the solution.

Here $\mathcal{L} = |R| \cdot c + |K| + |L| \cdot b - \lambda \cdot (b \cdot c - r)$. We consider the problem

$$\min_{b,c \geq 0} [|R| \cdot c + |K| + |L| \cdot b - \lambda \cdot (b \cdot c - r)]$$

We make derivatives of $\mathcal{L}$ with respect to variables b and c.

$$\frac{\partial \mathcal{L}}{\partial b} = |L| - \lambda \cdot c = 0 \Rightarrow |L| = \lambda \cdot c \ ; \qquad \frac{\partial \mathcal{L}}{\partial c} = |R| - \lambda \cdot b = 0 \Rightarrow |R| = \lambda \cdot b$$

We obtain the Lagrangian equations: $|L| \cdot b = \lambda \cdot r$, and $|R| \cdot c = \lambda \cdot r$
We can multiply these two equations together to get $|L| \cdot |R| = \lambda^2 \cdot r$
From here, we deduce $\lambda = \sqrt{|R| \cdot |L|/r}$
By substituting the value of λ in the Lagrangian equations, we get:

$$b = \sqrt{|R| \cdot r/|L|}, \text{ and } c = \sqrt{|L| \cdot r/|R|}$$

Then, from expression (11), we get the minimum communication cost of 3WJ

$$|R| \cdot \sqrt{|L| \cdot r/|R|} + |K| + |L| \cdot \sqrt{|R| \cdot r/|L|} \approx 2 \cdot |R| \cdot \sqrt{r}$$

Next, we specify the intermediate data size of the cascade of 2 two-way joins:

$$|R| \cdot |K| \cdot \alpha + |L| \approx |R|^2 \cdot \alpha \ (\text{where } |R| \cdot \alpha > 1)$$

The cost of 3WJ, $O(|R| \cdot \sqrt{r})$, is compared with the cost of the two-way join cascade $O(|R|^2 \cdot \alpha)$. We can conclude that 3WJ will be better than the cascade when $\sqrt{r} < |R| \cdot \alpha$. In other words, for 3WJ, there is a limit on the number of reducers $r < (|R| \cdot \alpha)^2$ and Theorem 4 is hence proved.   □

In general, we can extend Theorem 4 for 3WJ with n join key columns using an n-dimensional reducer matrix. For example, a 3WJ $R(A, B) \bowtie K(B, C) \bowtie L(C, A)$ with three join attributes A, B, and C. This three-way join needs a three-dimensional reducer matrix. The three-way join will become more efficient than a cascade of 2 two-way joins when $r < (|R| . \alpha)^3$ and its amount of communication is $3 \cdot |R| \cdot \sqrt[3]{r}$. In fact, choosing the number of reducers to satisfy this condition is not difficult. For example, if $|R| \cdot \alpha = 15$, as might be the case for the Web incidence matrix, we can choose the number of reducers r up to 3375. We can now characterize the cost of three-way join using filters.

Theorem 5. *A* 3WJ, *$R(A, B) \bowtie K(B, C) \bowtie L(C, D)$, is more efficient with filters than without filters when C_{pre} is negligible or less than the cost of processing non-matching data. Moreover, the* 3WJ *using the filters is also more efficient than the two-way join cascade using the filters when $r < (|R'| \cdot \alpha)^2$. With using the filters, the size of the intermediate data is defined by*

$$|D'| = \begin{cases} 2 \cdot |R'| \cdot \sqrt{r} & , \text{ for } 3\text{WJ using the filters.} \\ |R'|^2 \cdot \alpha & , \text{ for a cascade of 2 two-way joins using the filters.} \end{cases}$$

$$|R'| = \delta \cdot |R| + f_I \cdot (1 - \delta) \cdot |R|, R' \text{ is the filtered dataset of one input.}$$

where r is the number of reducers, α is the probability of two tuples from different datasets to match on the join key, $|R| = |K| = |L|$, δ is the ratio of the joined records of one input dataset with another, and f_I is the false intersection probability between the datasets.

Proof. Theorems 2 and 3 show that joins with the filters is more efficient than without the filters if C_{pre} is negligible or less than the cost of non-matching data. The following inequalities hold: $0 < \delta << 1$ and $0 < f_I << 1$

$$\Rightarrow \delta \cdot |R| + f_I \cdot (1 - \delta) \cdot |R| < |R| \Rightarrow |R'| < |R|$$

Combining this equality with Theorem 4, we can easily prove Theorem 5.   □

Chain Joins. Consider a chain join over n input datasets R_1, R_2, ..., R_n. We analyze OCJ-3WJ with the EIF filters presented in Sect. 3.2. The chain join is executed as a sequence of 3WJ jobs, $\ddot{J} = \{J_2, J_4, J_6, \ldots, J_{n-1}\}$. J_1 scans $R_1, \ldots, R_n$ inputs for building the filters. Each iteration carries out the join of three inputs, $R_{1,\ldots,2i-1}$, R_{2i}, and R_{2i+1}, where $1 \leq i \leq \lfloor (n-1)/2 \rfloor$. If n is even, OCJ-3WJ contains an additional two-way join job of $R_{1,\ldots,n-1}$ and R_n. We extend the cost model of 3WJ as follows:

$$C(\ddot{J}) = C_{pre} + \lceil (n-1)/2 \rceil \cdot C_{distCache} + C_{2wJoin}$$

$$+ \sum_{i=1}^{\lfloor (n-1)/2 \rfloor} (C_{read}(J_{2i}) + C_{sort}(J_{2i}) + C_{tr}(J_{2i}) + C_{write}(J_{2i}))$$

$$(12)$$

where:

- $C_{pre} = (\sum_{i=1}^{n} c_r \cdot |R_i|) + (c_l + c_t) \cdot m \cdot mp$
  - $\diamond$ $C_{pre} = 0$ and $m = 0$ for approaches without using filters.
  - $\diamond$ mp is the total number of map tasks.
- $C_{distCache} = 3 \cdot c_r \cdot m \cdot t$
  - $\diamond$ $C_{distCache} = 0$ for approaches without using filters.
- C_{2wJoin} is specified by Eq. (1), the cost of joining $R_{1,\ldots,n-1}$ and R_n.
  - $\diamond$ $C_{2wJoin} = 0$ if n is an odd number and greater than 2.
- $C_{read}(J_{2i}) = c_r \cdot |R_{1,\ldots,2i-1}| + c_r \cdot |R_{2i}| + c_r \cdot |R_{2i+1}|$
- $C_{sort}(J_{2i}) = c_l \cdot |D_i| \cdot 2 \cdot (\lceil \log_B |D_i| - \log_B(mp) \rceil + \lceil \log_B(mp) \rceil)$ [26]
  - $\diamond$ $|D_i|$ is the size of the intermediate data in the ith iteration.
- $C_{tr}(J_{2i}) = c_t \cdot |D_i|$
- $C_{write}(J_{2i}) = c_r \cdot |R_{1,\ldots,2i+1}| + a$
  - $\diamond$ $a = 2 \cdot c_r \cdot m \cdot t$, for building $BF(R_{1,\ldots,2i+1})$ in the ith iteration.
  - $\diamond$ $a = 0$, for $(2i + 1) = n$.

The computation of OCJ-2WJ is a sequence of $(n\text{-}1)$ two-way join jobs. This computation can be also considered as a sequence of $((n\text{-}1)/2)$ three-way join jobs in which each of them is executed by a cascade of 2 two-way join jobs. As a result, OCJ-2WJ has the extra costs of writing and re-reading the intermediate results of the two-way joins, and initializing additional jobs. On the other hand, OCJ-3WJ incurs the costs of data duplication to the reducers. From Theorem 5, we can show that OCJ-3WJ is more efficient than OCJ-2WJ when $r < (|R'| \cdot \alpha)^2$.

4.3   Recursive Joins

Cost Model. In the semi-naive algorithm, the number of iterations l is the longest path length in the relation graph minus 1, called the depth of the transitive closure. The first job J_1 reads K and $\Delta F_0 = F$, and caches K at the reducers. Each subsequent job J_i reads ΔF_{i-1} and scans partitions of K cached at the reducers (RIC). The dedup-diff job I_i reads the join output O_i containing duplicates, maps and shuffles tuples of O_i to the reducers in order to generate ΔF_i.

We base our analysis on the cost model introduced in [26] and adapt it to the evaluation of the recursive join. Table 3 gives the parameters.

The total cost of the recursive join is specified by:

$$
\begin{aligned}
C(\hat{J}) = C_K + \sum_{i=1}^{l} & C_{read}(J_i) + C_{sort}(J_i) + C_{tr}(J_i) + C_{cache}(J_i) + C_{write}(J_i) \\
+ \sum_{i=1}^{l} & C_{read}(I_i) + C_{sort}(I_i) + C_{tr}(I_i) + C_{cache}(I_i) + C_{write}(I_i)
\end{aligned}
\tag{13}
$$

where:

- $C_K = c_r \cdot |K| + c_l \cdot |K| \cdot 2 \cdot (\lceil \log_B |K| - \log_B(mp_K) \rceil + \lceil \log_B(mp_K) \rceil)$
  $+ (c_t + c_l) \cdot |K|$
- $C_{read}(J_i) = c_r \cdot |\Delta F_{i-1}|$
- $C_{sort}(J_i) = c_l \cdot |D_i| \cdot 2 \cdot (\lceil \log_B |D_i| - \log_B(mp_{\Delta Fi-1}) \rceil + \lceil \log_B(mp_{\Delta Fi-1}) \rceil)$
  [26]
- $C_{tr}(J_i) = c_t \cdot |D_i|$
- $C_{cache}(J_i) = c_l \cdot |K|$
- $C_{write}(J_i) = c_r \cdot |O_i|$
- $|D_i| = |\Delta F_{i-1}| = \beta_{i-1} \cdot |O_{i-1}|$
- $C_{read}(I_i) = c_r \cdot |O_i|$
- $C_{sort}(I_i) = c_l \cdot |D^+_i| \cdot 2 \cdot (\lceil \log_B |D^+_i| - \log_B(mp_{Oi}) \rceil + \lceil \log_B(mp_{Oi}) \rceil)$ [26]
- $C_{tr}(I_i) = c_t \cdot |D^+_i|$
- $C_{cache}(I_i) = c_l \cdot |D^+_i| \cdot (|F_{i-1}| \, / \, r) + c_l \cdot |\Delta F_i| \cdot (|F_{i-1}| \, / \, r + 1)$
- $C_{write}(I_i) = c_r \cdot |\Delta F_i|$
- $|D^+_i| = |O_i|$
- $|\Delta F_i| = \beta_i \cdot |O_i|$

The average size of the cache at each reducer is $(|F_{i-1}| \, / \, r)$. For each incoming tuple of O_i, the reducer probes the cache to get all tuples previously discovered. For each new tuple discovered, the reducer rewrites its entire cache along with the new tuple. Therefore, the total cost of accessing the cache in the dedup-diff job, $C_{cache}(I_i)$, includes the costs of reading the reducer cache for tuples of O_i and rewriting the reducer cache for new tuples of ΔF_i.

Cost Comparison. The total cost of REJ-FB is smaller than that of REJ-SHAW because the intermediate data of REJ-FB is less than that of REJ-SHAW ($|D'_i| < |D_i|$). The amount of intermediate data of REJ-FB is defined by:

$$
\begin{aligned}
|D'_i| &= \delta_K^{i-1} \cdot |\Delta F_{i-1}| + f(K) \cdot (1 - \delta_K^{i-1}) \cdot |\Delta F_{i-1}| \\
&= \delta_K^{i-1} \cdot |D_i| + f(K) \cdot (1 - \delta_K^{i-1}) \cdot |D_i| < |D_i|
\end{aligned}
\tag{14}
$$

Table 3. Parameters of our cost model for recursive joins

Parameter	Explanation
c_l	The cost of reading or writing data locally
c_r	The cost of reading/writing data remotely
c_t	The cost of transferring data from one node to another
$B+1$	The size of the sort buffer is $B+1$ pages (all costs are measured in seconds per page)
mp_K	The total number of map tasks of the dataset K
$mp_{\Delta F_{i-1}}$	The total number of map tasks of the incremental relation ΔF_{i-1}
mp_{Oi}	The number of the map tasks of the join output (O_i)
r	The number of reduce tasks
t	The number of tasktrackers
$|K|$	The size of the dataset K that is invariant in loops
$|\Delta F_{i-1}|$	The size of the incremental relation in the $(i-1)^{th}$ iteration ($|\Delta F_0| = |K|$)
$|\Delta F_i|$	The size of the incremental relation in the ith iteration. The dataset ΔF_i contains only the differences between the join output O_i and F_{i-1}
$|F_{i-1}|$	The size of all incremental relations in the iterations 0 to $i-1$ ($|\Delta F_0 \cup \cdots \cup \Delta F_{i-1}|$)
$|D_i|$	The intermediate data size of the join job J_i in the ith iteration
$|D^+_i|$	The intermediate data size of the dup-diff job I_i in the ith iteration
$|O_i|$	The size of the join processing output O_i. The output O_i may contain duplicate elements with F_{i-1} (previous incremental relations)
β_i	The difference ratio of the output O_i with F_{i-1}
C_K	The total cost to read, map and sort, shuffle, and cache K at the reducers (RIC) in the first iteration
$C_{read}(J_i)$	The total cost to read the incremental relation ΔF_{i-1} from DFS
$C_{sort}(J_i)$	The total cost to perform the sorting and copying of the join job at the map and reduce nodes
$C_{tr}(J_i)$	The total cost to transfer intermediate data of the join job among nodes
$C_{cache}(J_i)$	The total cost to locally read partitions of K cached at the reducers
$C_{write}(J_i)$	The total cost to write O_i to DFS
$C_{read}(I_i)$	The total cost to read the join output O_i from DFS
$C_{sort}(I_i)$	The total cost to perform the sorting and copying of the dup-diff job at the map and reduce nodes
$C_{tr}(I_i)$	The total cost to transfer intermediate data of the dup-diff job ($=O_i$) among the nodes
$C_{cache}(I_i)$	The total cost to locally read partitions of F_{i-1} cached at the reducers
$C_{write}(I_i)$	The total cost to write ΔF_i to DFS

where:

- δ_K^{i-1} is the ratio of the joined records of ΔF_{i-1} with K
- $f(K)$ is the false positive probability of the Bloom filter $BF(K.z)$

We need a pre-processing job for building the Bloom filter $BF(K.z)$ that is used in all iterations. The additional overhead of building the filter $BF(K.z)$ is:

$$C'_K = C_K + C_{pre} \tag{15}$$

where:

- $C_{pre} = c_r \cdot |K| + (c_l + c_t) \cdot m_k \cdot mp_k + c_r \cdot m_k \cdot t$
- m_k is the compressed size of the Bloom filter of the input dataset K (bits). It is the product of the size of the filter and the file compression ratio. If the size of the filter is small, the file compression ratio should be one.

Besides, on each iteration, the program also re-computes the global filter $BF(\Delta F_i.z)$ generated in the reduce phase of the dedup-diff job. The overhead of creating the filter $BF(\Delta F_i.z)$ is:

$$C'_{write}(I_i) = C_{write}(I_i) + (2 \cdot c_r \cdot m_{\Delta Fi} \cdot r + c_r \cdot m_{\Delta Fi}) \tag{16}$$

where:

- $m_{\Delta Fi}$ is the compressed size of the Bloom filter of the incremental dataset ΔF_i (bits)

Since the size of the filters is small, these extra overheads are negligible compared to the overheads associated with redundant data in the incremental dataset.

5   Experimental Evaluation for Filter-Based Equijoins

In this section, we present experimental results obtained from the execution of two-way joins, chain joins, and recursive joins.

5.1   Two-Way Joins

Cluster Environment and Datasets. All experiments were run on a cluster of 15 virtual machines using Virtualbox [28]. Each machine has two 2.4 Ghz AMD Opteron CPUs with 2 MB cache, 10 GB RAM and 100 GB SATA disks. The operating system is 64-bit Ubuntu server 12.04, and the java version is 1.7.0.21. We installed Hadoop [6] version 1.0.4 on all nodes. One of the nodes was selected to act as Master and ran the NameNode and the JobTracker processes; the remaining nodes host the TaskTrackers in charge of data storage and data processing. Each TaskTracker node was configured to run up to two simultaneous map tasks and two reduce tasks. The HDFS block size was set to 128MB, size of read/write buffer was 128 KB, heap-size for JVMs was set to 2048 M, and the number of reduce tasks set to 28.

All test datasets were produced by a data generation script of the Purdue MapReduce Benchmarks Suite [4], called "PUMA" which represents a broad

Table 4. Input datasets

Inputs	Test 1		Test 2		Test 3	
	size	*records*	*size*	*records*	*size*	*records*
Dataset1	15 GB	40,259,163	35 GB	92,681,333	55 GB	145,099,559
Dataset2	15 GB	40,108,215	35 GB	92,524,495	55 GB	139,573,823
Total	30 GB	80,367,378	70 GB	185,205,828	110 GB	284,673,382

range of MapReduce applications with high/low computing requirements and high/low shuffle volumes. The maximum number of columns in the datasets is 39 and string length in each column is set to 19 characters. The first column of *Dataset1* is a foreign key that refers to the fifth column of *Dataset2*. We used three test sets *Test 1*, *Test 2*, and *Test 3* with respective sizes 30 GB, 70 GB, and 110 GB. Table 4 summarizes the dataset sizes used in our experiments. The ratios of the joined records are 0.054 % (Test 1), 0.057 %(Test 2), 0.063 %(Test 3).

We executed our algorithm for the following join query.

```
SELECT *
FROM dataset1(column0..column20) d1, dataset2(column0..column20) d2
WHERE  d1.column0 = d2.column5
```

We particularly investigate four aspects: the number of intermediate tuples generated, the total execution time, the tasks timeline, and the scalability measured by varying the input size.

Evaluation of Approaches. In order to execute the filter-based algorithms efficiently, we specified the size of filters according to the cardinality of the join key values of datasets and chose the largest filter. There is a tradeoff between m and the probability of a false positive. Hence, the probability of a false positive f is approximated by:

$$f \approx \left(1 - e^{-\rho \cdot n/m}\right)^{\rho}$$

For a given false positive probability f, the size of the Bloom filter m is proportional to the number of elements n in the filter as shown in Table 5.

Table 5. Parameters for filters

Tests	f	ρ	n	m/n	m (bit)
Test 1	0.001	7	14,866	15	222,990
Test 2	0.0001	8	15,790	21	331,590
Test 3	0.0001	8	15,790	21	331,590

where ρ is the number of hash functions, and m/n is the number of bits allocated to each join key.

We can determine optimized parameters for the filter (e.g. f, ρ and m) [10]. In practice, however, we should choose values less than an optimized value to reduce computational overhead. As shown in Table 5, we deliberately select various values of f, ρ and m/n for the experiments to consider if they might affect our join performance. The filter files generated in the tests are compressed with gzip.

Table 6. Number of intermediate tuples (Map output)

Join algorithms	Test 1 (30 GB)	Test 2 (70 GB)	Test 3 (110 GB)
IFBJ	43,453	106,116	179,091
BJ	40,276,915	92,747,151	145,206,430
RSJ	80,320,684	185,098,062	284,510,488

The intermediate data size (Map output) is given in Table 6. The Reduce-side join (without filter) is the most inefficient solution, although it runs as a single job. This is correlated to the large size of intermediate data. Note that the number of intermediate tuples generated in this case is almost equal to the number of Map input records, see Tables 4 and 6. This slight difference is because a few tuples of *Dataset2* have less than 6 columns, and so they have been eliminated.

Filter-based joins are more efficient in general. BJ and IFBJ include the pre-processing job and the filtering operation to improve the join performance.

The number of intermediate tuples produced by BJ is considerably reduced with respect to RSJ. However, in comparison to IFBJ (see in Table 6), BJ still produces much more intermediate data because the filtering operation is only executed on one input dataset (*Dataset1*). This situation is overcome by IFBJ.

Looking at BJ and IFBJ, Table 6 points out that BJ generates more intermediate data than IFBJ. Namely, for the 110 GB test, BJ produces 145,206,430 intermediate tuples, whereas IFBJ produces 179,091 tuples. The experiments reported above are consistent with our theoretical analysis (Theorem 1).

Next, we evaluate the efficiency of these join algorithms by comparing the total execution time. As a general fact, the join algorithms generating less intermediate data turn out to be faster, even if we sum up the cost of the pre-processing and join jobs.

Table 7 gives the total execution time of the pre-processing job and the join job for each algorithm. Regarding pre-processing, the cost of the filter-based joins is related to the size of the data accessed to build the filter(s). In particular, IFBJ has to scan two input datasets. However, it pays off, since once the filters are available, the cost of join jobs is drastically reduced.

Figure 8 demonstrates that the best execution results from using intersection filters. Their total execution time is significantly reduced compared to BJ in spite of the time spent in the pre-processing job. The total execution time of

Table 7. Execution of pre-processing job and join job (in minutes)

Joins	Test 1 (30 GB)			Test 2 (70 GB)			Test 3 (110 GB)		
	Pre-proc.	Join job	Total time	Pre-proc.	Join job	Total time	Pre-proc.	Join job	Total time
IFBJ	3.17	6.15	9.32	6.45	24.25	31.10	10.00	92.12	102.12
BJ	2.12	17.07	19.19	3.63	43.63	47.26	5.22	139.58	145.20
RSJ	0	28.25	28.25	0	70.13	70.13	0	150.00	150.00

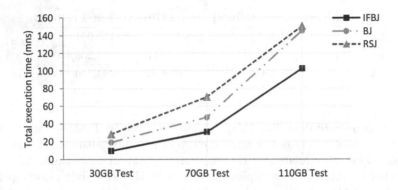

Fig. 8. Total execution time

IFBJ increases from about 10 to 105 (mns), whereas that of BJ ranges from 19.19 to 145.20 (mns). The worst execution is RSJ, ranging 28.25 to 150 (mns). The smaller cost of IFBJ compared to the others (Table 7), is analyzed in Theorem 2.

Finally, we analyze the sequence of tasks during job execution (called task timelines). We do not examine the task timelines of the pre-processing job which is negligible compared to the join query over large datasets (see Table 7).

Figure 9 represents the task timelines of 70GB join jobs. These graphs are created by parsing log files generated by Hadoop during the job execution (555 map tasks and 28 reduce task, processing 185,205,828 input records and producing 26,062,967 output records). Each graph shows the respective timelines for map, shuffle and reduce phases.

There is a notable difference between the task timeline of IFBJ and that of other joins. The execution time of all map and reduce tasks of IFBJ, Fig. 9(a), is significantly reduced compared to BJ and RSJ, Fig. 9(b) and (c). Besides, the map and reduce phases of IFBJ finished earlier than BJ and RSJ because they produce less intermediate data and, as a consequence, the total cost of the local I/O, sort, and remote data copy is also smaller. Joins that use the intersection filter are the most efficient solutions because of their better data filtering efficiency.

The efficiency of filter-based joins depends on the ratio of non-matching tuples. The threshold is defined by the two parameters $\delta_{dataset2}$ and $\delta_{dataset1}$, which are the ratios of matching tuples. Figure 10 shows the execution time

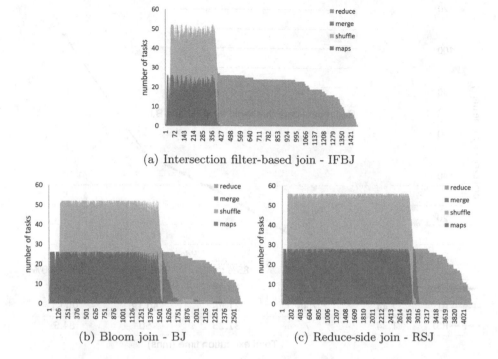

(a) Intersection filter-based join - IFBJ

(b) Bloom join - BJ   (c) Reduce-side join - RSJ

Fig. 9. 70GB Task timelines during the execution of the join job

of algorithms for several values of these parameters, in order to identify their impact.

We start with an extreme case (first column) where domain of join attributes in *Dataset1* and *Dataset2* are disjoint. A IFBJ is then able to discover the empty intersection and the join job can therefore be omitted altogether, and their costs represents only that of the pre-processing job. This cost is roughly comparable with that of RSJ because of the small size of the dataset which make the join job fast enough. Filtered joins should not be used for small input datasets because the cost of building and broadcasting filter(s) becomes relatively significant.

We next examine the cases of a high ratio of matching tuples (85 % : 4 %) and (95 % : 65 %). They represent respectively the thresholds for filter-based join resulting from our analysis. Figure 10 clearly shows that this is the point where filters become counter-productive. This can be determined at compile time based on the ratios $\delta_{dataset2} : \delta_{dataset1}$.

The last case shows a join between fully matching datasets (100 % : 100 %), in which case RSJ is the best solution.

5.2 Multi-way Joins

Cluster Environment and Datasets. We run experiments for the chain join on another computer cluster of 15 virtual machines using KVM (Kernel-based

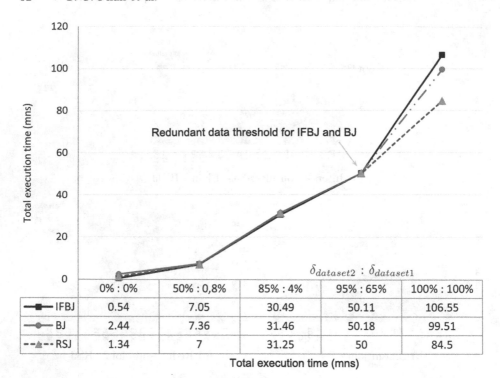

Fig. 10. Identification of threshold for non-matching tuples for joins with 2 GB inputs

Virtual Machine) [18]. Each machine has two 1.4 Ghz AMD Opteron CP Us with 512 KB cache, 5 GB RAM and 100 GB SATA disks. We installed Hadoop [6] version 1.0.4 on all nodes. The other configurations of this cluster are similar to the ones of the cluster running the experiments of the two-way joins. The number of reduce tasks is set to 25.

All datasets were also produced by the data generation script of the PUMA. The maximum number of columns in the datasets is 39 and string length in each column is set 19 characters. The datasets *Dataset1*, *Dataset2*, *Dataset3*, and *Dataset4* contain the join key columns *column1* (c_1), *column2* (c_2), *column3* (c_3), and *column4* (c_4). Tables 8 and 9 summarizes the different input dataset sizes and the joined record ratios, resp.

The chain join algorithms developed in our experiments are the Reduce-side join cascade (CJ-RSJ), the Bloom join cascade (CJ-BJ), the IF-based join cascade (CJ-IFBJ), the optimized two-way join cascade (OCJ-2WJ), and the optimized three-way join cascade (OCJ-3WJ). The following chain join query is used.

```
SELECT * FROM  dataset1(c1..c10) d1, dataset2(c1..c10) d2,
               dataset3(c1..c10) d3, dataset4(c1..c10) d4
WHERE  d1.c2 = d2.c2 AND d2.c3 = d3.c3 AND d3.4 = d4.c4
```

Table 8. Input datasets used in three tests

Inputs	Test 1		Test 2		Test 3	
	size	*records*	*size*	*records*	*size*	*records*
dataset1	10 GB	26,836,497	20 GB	53,675,946	20 GB	53,682,929
dataset2	3 GB	8,051,454	10 GB	26,838,960	30 GB	73,881,305
dataset3	10 GB	26,836,497	20 GB	53,675,946	20 GB	53,682,929
dataset4	3 GB	8,051,454	10 GB	26,838,960	30 GB	73,881,305
Total	26 GB	69,775,902	60 GB	161,029,812	100 GB	255,128,468

Table 9. The ratios of the joined records of the datasets (%)

Inputs	Test 1	Test 2	Test 3
dataset1	0.721722639	0.304090688	0.123521020
dataset2	0.216530370	0.152050936	0.169996205
dataset3	0.721722639	0.304090688	0.123521020
dataset4	0.216530370	0.152050936	0.169996205

Evaluation. The experiments use the parameters of the Boom filters given in Table 10.

In order to confirm the cost model of chain joins (Sect. 4.2), we first examine the amount of intermediate data (Table 11)

Table 11 shows that CJ-RSJ and CJ-BJ generate much more intermediate data than any algorithms using the (extended) intersection filters. Figure 11 helps us

Table 10. Parameters of Bloom filters

Tests	f	ρ	n	m/n	m (bit)
Test 1	0.000101	8	13,147	21	276,087
Test 2	0.000101	8	13,840	21	290,640
Test 3	0.000101	8	15,295	21	321,195

Table 11. Number of intermediate tuples (all map outputs)

Chain join algorithms	Test 1 (26 GB)	Test 2 (60 GB)	Test 3 (100 GB)
CJ-IFBJ	1,309,349	1,469,048	1,497,692
CJ-BJ	45,402,907	89,201,979	89,248,190
CJ-RSJ	88,296,034	196,465,292	290,582,143
OCJ-2WJ	1,281,036	1,417,684	1,445,428
OCJ-3WJ	1,221,769	1,359,575	1,385,053

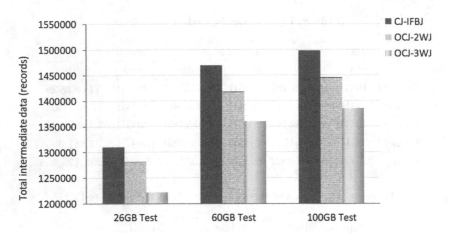

Fig. 11. Total intermediate data of the chain join

to obtain a visual comparison of the intersection filter-based chain joins. OCJ-3WJ has the least amount of intermediate data because it has the least number of jobs, and filters out almost all non-matching tuples in intermediate results. The intermediate data amount of OCJ-2WJ is slightly greater than the intermediate data amount of OCJ-3WJ, as analyzed by Theorem 5. However, OCJ-2WJ is still better than CJ-IFBJ chain joins which do not fully prevent non-matching tuples to propagate throughout the join chain.

Next, we examine the total output of the chain join algorithms (Fig. 12). The total output consists of all the intermediate data generated in the map phase

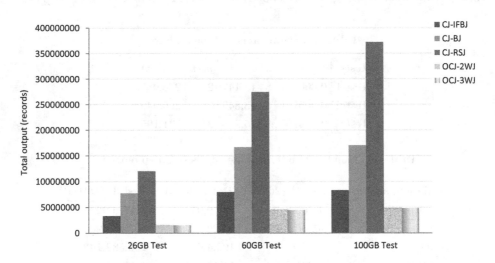

Fig. 12. Total output data (Map output + Reduce output)

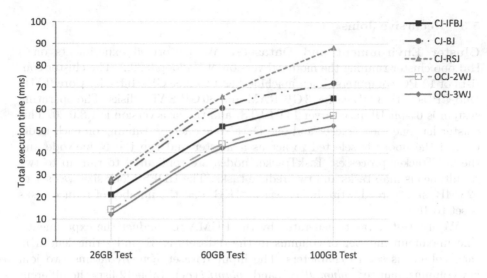

Fig. 13. Total execution time

and the intermediate join results. In other words, it includes all map output tuples and reduce output tuples produced during the chain join.

As shown in Fig. 12, CJ-RSJ and CJ-BJ generate the largest outputs; whilst the OCJ joins (e.g. OCJ-2WJ and OCJ-3WJ) using the extended intersection filters produce the least output. The CJ-IFBJ joins generally produce a little more output than the OCJ joins. The main reason is that the OCJ joins have the ability to filter out much more non-matching tuples than the others.

Both CJ-RSJ and CJ-BJ exhibit a similar pattern, with a significant cost increase from 26GB to 100GB. Obviously, CJ-RSJ has the highest cost with 119,928,957 tuples for Test 1, (77,035,830 for CJ-BJ and 32,942,272 for the CJ-IFBJ joins). This is even worse with Test 3, CJ-RSJ produces 371,782,345 tuples compared to 170,448,392 for CJ-BJ and 82,697,894 for the CJ-IFBJ joins.

Let us finally discuss the performance comparison, summarized by Fig. 13. The run time is clearly correlated to the size of the intermediate data, as confirmed by the comparison of the relative performance of the algorithms and the number of tuples shipped during the execution of joins.

The two bottom graphs show the total execution times of the OCJ joins (OCJ-3WJ and OCJ-2WJ), the next three ones deal with CJ-IFBJ, and the two top graphs show CJ-BJ and CJ-RSJ. For the largest dataset (100GB), OCJ-3WJ and OCJ-2WJ run time is about 52.57 and 57.22 min respectively, while the CJ-IFBJ joins run time is about 65.13 min. CJ-BJ and CJ-RSJ run time is much longer, about 72.09 and 88.34 min resp. This shows the high benefit of filtering out useless data, as carrying this data all over the process constitutes a strong penalty. Note that these costs include the pre-processing step for filter-based joins. In a scenario where filters are built once, and the joins processed many times, the benefit of the approach is even reinforced. The results of these experiments are consistent with our cost analysis presented in Sect. 4.2.

5.3 Recursive Joins

Cluster Environment and Datasets. We performed experiments on a HaLoop cluster running the modified version of Hadoop 0.20.2. The cluster consists of 12 PC computers. Each machine has two 2.53 GHz Intel(R) Core(TM)2 Duo CPUs with 3 MB cache, 3 GB RAM and 80 GB SATA disks. The operating system is 64-bit Ubuntu server 14.04 LTS, and the java version is 1.8.0.20. This cluster has one TaskTracker and one DataNode daemon running on each node. One of the nodes is selected to act as a master and run the NameNode and the JobTracker processes. TaskTracker nodes are configured to run up to two simultaneous map tasks and two reduce tasks. The HDFS block size was set to 128 MB, size of read/write buffer was 128 KB, and the number of reduce tasks is set to 16.

We use test datasets generated by the PUMA to conduct the experiments. The maximum number of columns in the datasets is 31 and string length in each column is set 19 characters. The input dataset *Know* contains two join key columns, namely, $column0\,(c_0)$, and $column1\,(c_1)$. Table 12 lists the different sizes of the dataset *Know* used in our tests.

Table 12. Input dataset *Know* with different sizes

Test	Size	Records
Test 1	10 GB	53,674,078
Test 2	20 GB	107,349,426
Test 3	30 GB	150,000,054

The following recursive join query is used to evaluate our experiments.

$$Friend(c_0, c_1, \ldots, c_{30}) \longleftarrow Know(c_0, c_1, \ldots, c_{30})$$
$$Friend(c_0, c_1, \ldots, c_{30}) \longleftarrow Friend(c_0, c_1, \ldots, c_{30}) \bowtie_{c1=c'0} Know(c'_0, c'_1, \ldots, c'_{30})$$

Evaluation. The filters' parameters used in the filter-based approach are listed in Table 13.

Table 13. Parameters of filters

Tests	f	ρ	n	m/n	m (bit)
Test 1	0.000101	8	7,111	21	149,331
Test 2	0.000101	8	7,123	21	149,583
Test 3	0.000101	8	7,130	21	149,730

Table 14. The total number of intermediate tuples

Recursive join approaches	Test 1 (10 GB)	Test 2 (20 GB)	Test 3 (30 GB)
REJ-SHAW	215,609,705	431,589,879	602,707,978
REJ-FB	188,597,706	377,403,437	527,220,188

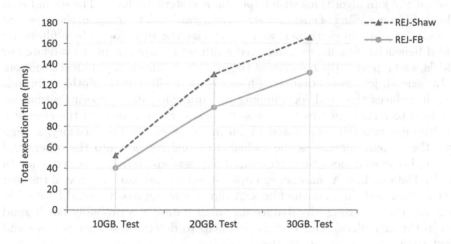

Fig. 14. Total execution time

We first examine the total map output (Table 14). The Shaw's approach (REJ-SHAW) generates more intermediate data than the filter-based approach (REJ-FB). For the tests from 10 GB to 30 GB, REJ-SHAW generates from 215,609,705 to 602,707,978 tuples, whilst REJ-FB has less than from 188,597,706 to 527,220,188 respectively. This is because the intermediate data of the join jobs in REJ-SHAW contains a lot of non-matching tuples, whereas REJ-FB uses the intersection filter to eliminate these non-matching tuples from the intermediate data of the join jobs.

Next, we examine the efficiency of the recursive join approaches. The total execution time of REJ-SHAW is compared to that of REJ-FB. Let us look in Fig. 14 for more details.

Figure 14 presents the total execution time of the pre-processing job and the iterative (join + dedup-diff) jobs for each algorithm. The cost of REJ-FB is considerably reduced in spite of the additional pre-processing job.

With the 10 GB input dataset *Know*, the total execution time of the Shaw's approach is higher than that of REJ-FB. This remains so through the other tests.

6 Conclusions and Future Work

The join operation is one of the essential operations for data analysis. Join evaluation is expensive and not straightforward to implement with MapReduce. This paper makes three contributions. First, we attempt to gather in a uniform setting some of the main approaches recently proposed for the most common types of joins. In particular, we systematically considered the introduction of filters in execution plans. Filters are known to greatly improve the cost of distributed joins thanks to their ability to avoid network transfer of useless data. We showed how to adapt the join algorithms with filters, on a systematic basis. The second contribution is a modeling of cost that serves as a yardstick to compare the expected efficiency of joins. In particular, we characterize the situations where filters are indeed beneficial. Finally, we conducted a full set of experiments to validate our models, and reported the behavior of the join algorithms in a practical situation.

In general, join evaluation using filters is more efficient than other solutions since it reduces the need for shipping non-matching data. Specific situations may lead to reconsider this general assumption. For instance, in the case of a join between two relations linked by an integrity constraint (primary, foreign key), the system guarantees the inclusion of one key set into the other, and filtering becomes useless. Such structured datasets are arguably not common in the Big Data realms. As another example, small dataset size may reveal the cost of producing and shipping the filters. A direct join approach should be used in that case (in fact using MapReduce for small datasets is probably not a good idea in the first place). Our cost models help to detect those special cases and adopt the proper evaluation strategy.

The present study could be extended in several directions. First, a complete coverage would include star joins, and in general joins amongst n relations linked by complex relationships. Given the complexity of matching such a general setting with a MapReduce framework, we consider that the set of joins cases investigated in what precedes constitute a satisfying set of primitives to start with. Regarding our experimental evaluation, we did our best to use the state-of-the-art MapReduce framework (e.g., HaLoop). We note that some recent distributed engines (e.g., Spark [7], Stratosphere/Flink [5,34]) natively bring some of the features examined here, and notably iterations. At a physical level, they support caching of intermediate result, if possible in main memory. This strengthens our expectation that joins (including recursive joins) as studied here, constitute the basic building block of sophisticated algorithms for machine learning and data mining, which stand as the most promising outcome of Big Data processing in a near future. In this respect, the present study stands as a first step toward the design of an optimizer for distributed query processing, apt at considering complex integration of iterative, recursive and multi-set operators. We plan to investigate in the future the foundations of such an optimizer.

References

1. Afrati, F.N., Borkar, V., Carey, M., Polyzotis, N., Ullman, J.D.: Cluster computing, recursion and datalog. In: de Moor, O., Gottlob, G., Furche, T., Sellers, A. (eds.) Datalog 2010. LNCS, vol. 6702, pp. 120–144. Springer, Heidelberg (2011)

2. Afrati, F.N., Borkar, V.R., Carey, M.J., Polyzotis, N., Ullman, J.D.: Map-reduce extensions and recursive queries. In: Proceedings of the International Conference on Extending Database Technology (EDBT), Uppsala, Sweden, pp. 1–8 (2011)
3. Afrati, F.N., Ullman, J.D.: Optimizing joins in a map-reduce environment. In: Proceedings of the International Conference on Extending Database Technology (EDBT), Lausanne, Switzerland, pp. 99–110 (2010)
4. Ahmad, F.: Puma benchmarks and dataset downloads (2012). https://engineering.purdue.edu/~puma/datasets.htm. Accessed: 18 June 2015
5. Apache: Flink. http://flink.apache.org. Accessed: 18 June 2015
6. Apache: Hadoop. http://hadoop.apache.org/. Accessed: 18 June 2015
7. Apache: Spark. https://spark.apache.org. Accessed: 18 June 2015
8. Blanas, S., Patel, J.M., Ercegovac, V., Rao, J., Shekita, E.J., Tian, Y.: A comparison of join algorithms for log processing in mapreduce. In: Proceedings of the 2010 ACM SIGMOD International Conference on Management of Data, SIGMOD 2010, pp. 975–986. ACM, New York (2010)
9. Bloom, B.H.: Space/time trade-offs in hash coding with allowable errors. Commun. ACM **13**(7), 422–426 (1970)
10. Broder, A.Z., Mitzenmacher, M.: Survey: network applications of Bloom filters: a survey. Internet Math. **1**(4), 485–509 (2003)
11. Bruno, N., Kwon, Y., Wu, M.C.: Advanced join strategies for large-scale distributed computation. Proc. VLDB Endow. **7**(13), 1484–1495 (2014)
12. Bu, Y., Howe, B., Balazinska, M., Ernst, M.D.: The HaLoop approach to large-scale iterative data analysis. VLDBJ **21**(2), 169–190 (2012)
13. Dean, J., Ghemawat, S.: MapReduce: simplified data processing on large clusters. In: Proceedings of the International Symposium on Operating System Design and Implementation (OSDI), San Francisco, California, pp. 137–150 (2004)
14. Doulkeridis, C., Nrvg, K.: A survey of large-scale analytical query processing in mapreduce. VLDB J. **23**(3), 355–380 (2014)
15. Facebook,: Facebook reports fourth quarter and full year 2013 results - facebook (2014). http://investor.fb.com/releasedetail.cfm?ReleaseID=821954. Accessed: 18 June 2015
16. Hassan, M.A.H., Bamha, M.: Semi-join computation on distributed file systems using map-reduce-merge model. In: Proceedings of the Symposium on Applied Computing (SAC), Sierre, Switzerland, pp. 406–413 (2010)
17. Idreos, S., Liarou, E., Koubarakis, M.: Continuous multi-way joins over distributed hash tables. In: Proceedings of the EDBT, Nantes, France, pp. 594–605 (2008)
18. KVM: Kernel virtual machine. http://www.linux-kvm.org/page/Main_Page. Accessed: 18 June 2015
19. Lam, C.: Hadoop in Action. Manning Publications, Greenwich (2010)
20. Lee, K.H., Lee, Y.J., Choi, H., Chung, Y.D., Moon, B.: Parallel data processing with mapreduce: a survey. SIGMOD Rec. **40**(4), 11–20 (2012)
21. Lee, T., Im, D.H., Kim, H., Kim, H.J.: Application of filters to multiway joins in MapReduce. Math. Probl. Eng. **2014**, 11 (2014)
22. Lee, T., Kim, K., Kim, H.J.: Join processing using Bloom filter in MapReduce. In: Proceedings of the RACS, San Antonio, TX, USA, pp. 100–105 (2012)
23. Lee, T., Kim, K., Kim, H.J.: Exploiting bloom filters for efficient joins in MapReduce. Inf. Int. Interdisc. J. **16**(8), 5869–5885 (2013)
24. Li, F., Ooi, B.C., Özsu, M.T., Wu, S.: Distributed data management using MapReduce. ACM Comput. Surv. **46**(3), 31:1–31:42 (2014)

25. Liu, L., Yin, J., Gao, L.: Efficient social network data query processing on MapReduce. In: Proceedings of the Workshop on HotPlanet, Hong Kong, China, pp. 27–32 (2013)
26. Nykiel, T., Potamias, M., Mishra, C., Kollios, G., Koudas, N.: MRShare: sharing across multiple queries in MapReduce. Proc. Very Large Data Bases Endowment (PVLDB) 3(1), 494–505 (2010)
27. Okcan, A., Riedewald, M.: Processing theta-joins using mapreduce. In: Proceedings of the 2011 ACM SIGMOD International Conference on Management of Data, SIGMOD 2011, pp. 949–960. ACM, New York (2011)
28. Oracle: Oracle vm virtualbox. https://www.virtualbox.org. Accessed: 18 June 2015
29. Ordonez, C.: Optimizing recursive queries in SQL. In: Proceedings of the SIGMOD International Conference on Management of Data, Baltimore, Maryland, USA, pp. 834–839 (2005)
30. Phan, T.C., d'Orazio, L., Rigaux, P.: Toward intersection filter-based optimization for joins in mapreduce. In: Proceedings of the 2nd International Workshop on Cloud Intelligence, Cloud-I 2013, pp. 2:1–2:8. ACM, New York (2013)
31. Sakr, S., Liu, A., Batista, D., Alomari, M.: A survey of large scale data management approaches in cloud environments. IEEE Commun. Surv. Tutorials 13(3), 311–336 (2011)
32. Sakr, S., Liu, A., Fayoumi, A.G.: The family of mapreduce and large-scale data processing systems. ACM Comput. Surv. 46(1), 11:1–11:44 (2013)
33. Shaw, M., Koutris, P., Howe, B., Suciu, D.: Optimizing large-scale Semi-Naive Datalog evaluation in Hadoop. In: Proceedings of the International Workshop on Datalog 2.0 (Datalog), Vienna, Austria, pp. 165–176 (2012)
34. Stratosphere: Next generation big data analytics platform. http://stratosphere.eu. Accessed: 18 June 2015
35. Tan, K.L., Lu, H.: a note on the strategy space of multiway join query optimization problem in parallel systems. SIGMOD Rec. 20(4), 81–82 (1991)
36. Ullman, J.D.: Principles of Database and Knowledge-Base Systems, vol. I. Computer Science Press, Rockville (1988)
37. White, T.: Hadoop: The Definitive Guide. O'Reilly, Sebastopol (2012)
38. Zhang, C., Li, J., Wu, L., Lin, M., Liu, W.: Sej: an even approach to multiway theta-joins using mapreduce. In: CGC 2012, pp. 73–80. IEEE Computer Society (2012)
39. Zhang, C., Wu, L., Li, J.: Optimizing distributed joins with bloom filters using MapReduce. In: Kim, T., Cho, H., Gervasi, O., Yau, S.S. (eds.) GDC, IESH and CGAG 2012. CCIS, vol. 351, pp. 88–95. Springer, Heidelberg (2012)
40. Zhang, C., Wu, L., Li, J.: Efficient processing distributed joins with bloom filter using mapreduce. Int. J. Grid Distrib. Comput. (IJGDC) 6(3), 43–58 (2013)
41. Zhang, X., Chen, L., Wang, M.: Efficient multi-way theta-join processing using mapreduce. Proc. VLDB Endow. 5(11), 1184–1195 (2012)

A Constraint Optimization Method for Large-Scale Distributed View Selection

Imene Mami$^{(\boxtimes)}$, Zohra Bellahsene, and Remi Coletta

University Montpellier 2, LIRMM, Montpellier, France
{mami,bella,coletta}@lirmm.fr

Abstract. View materialization is a commonly used technique in many data-intensive systems to improve the query performance. Increasing need for large-scale data processing has led to investigating the view selection problem in distributed complex scenarios where a set of cooperating computer nodes may share data and issue numerous queries. In our work, the view selection and data placement problem is studied given a limited amount of resources e.g. storage space capacity per computer node and maximum view maintenance cost. We also consider the IO and CPU costs for each computer node as well as the network bandwidth. To address this problem, we have proposed a constraint programming approach which is based on constraint reasoning to tackle problems that aim to satisfy a set of constraints. Then, we have designed a set of efficient heuristics that result in a drastic reduction in the solution space so that the problem becomes solvable for complex scenarios consisting of realistically large numbers of sites, queries and views. Our experimental study shows that our approach performs consistently better compared to a practical approach designed for large-scale distributed environments which uses a genetic algorithm to compute which view has to be materialized at what computer node.

Keywords: Distributed database design · Modeling and management · Query processing and optimization · Materialized views · Constraint optimization problem

1 Introduction

Materialized views have long been used in many data-intensive systems, such as commercial or scientific database systems, to obtain significant performance improvements when processing complex queries. View materialization would improve query evaluation by avoiding re-computation of expensive query operations. User queries can be answered using the information stored at the view relations and need not to be translated and shipped to the original sources for execution. Indeed, data sources may contain several millions of tuples. Therefore, scanning these data sources can take a significant amount of time. Because optimal set of materialized views can significantly speed up query processing, research has provided a huge set of view-based techniques for the efficient query

© Springer-Verlag Berlin Heidelberg 2016
A. Hameurlain et al. (Eds.): TLDKS XXV, LNCS 9620, pp. 71–108, 2016.
DOI: 10.1007/978-3-662-49534-6_3

evaluation but most of these works presents a solution in the central case. The growing interest in large-scale data analytics has turned query optimization problems in distributed data-intensive systems to a challenging and critical task. Such problems are typical for large retailer companies. In such complex scenarios, multiple computer nodes with different resource constraints (e.g., CPU, IO, storage space capacity, network bandwidth, etc.) query and update numerous base relations on different sites (i.e., computer nodes). A key factor to ensure query performance in such context is the intelligent placement of materialized views at different sites on the network.

Consider a simple example showing how materialized views can improve query evaluation. Let us consider the SQL query q_1 (defined over the TPC-H schema [2]) which finds the minimal supply cost for each product supplied in the USA.

q_1:**Select PS.partkey, Min(PS.supplycost)**
  From PartSupp PS, Supplier S, Nation N
  Where PS.suppkey=S.suppkey and S.nationkey=N.nationkey and
  N.name='USA'
  Group By PS.partkey;

Note that the base relations PartSupp (PS), Supplier (S) and Nation (N), which contain very large numbers of tuples, are stored at site s_2 and the query q_1 is posed frequently at site s_1. Suppose that two sites are connected by a slow connection speed. Therefore, the cost of transferring large amounts of data from s_2 to s_1 is high. This communication cost has to be paid every time the site s_1 issues the query q_1. One way to reduce the evaluation cost of the query q_1, is to use pre-computed results (i.e., the result of query q_1 or part of it), in the form of views over data. For instance, answering the query q_1 using the view v_1 (see below $q_{1_{v_1}}$). The view v_1 stores the list of suppliers who are from USA and was selected to be materialized at s_1. The costs of performing subsequent queries would be reduced significantly by using in their evaluation the materialized view v_1.

v_1:**CREATE MATERIALIZED VIEW** $v1$ **AS**
  Select *
  From Supplier S, Nation N
  Where S.nationkey=N.nationkey and N.name='USA';

$q_{1_{v_1}}$:**PS.partkey, Min(PS.supplycost)**
  From v_1**, PartSupp PS**
  Where v_1**.suppkey=PS.suppkey**
  Group By PS.partkey;

Obviously, materialized views have significantly improved the query performance. However, the result of the queries may be too large to fit in the available storage space at a specific site. Another concern is the cost of view maintenance. Whenever the data at the sources are changed, the materialized views built on it have to be updated in order to compute up-to-date query results. Indeed, the cost of view maintenance may offset the performance advantages provided by the view materialization. Thus, the view is considered as beneficial if and only if its materialization reduces significantly the query processing cost without increasing significantly the view maintenance cost. On the other hand, it is possible that some other neighbors sites, who have stored (part of) the required data, can be accessed through a much faster network connection, compared to the sites which provides original content. This highlights the need of considering resource constraints (e.g., storage space capacity of each site, view maintenance time and network bandwidth) while making view selection and data placement decisions. The heterogeneity between the different sites of the network (e.g., sites with different constraints on CPU and I/O) has also to be considered during the view selection and data placement process. The problem of deciding which views have to be selected and at what sites should be materialized given a limited amount of resources, is referred as the view selection problem and is known to be a NP-complete optimization problem. To the best of our knowledge, no past work described in the open literature has addressed this problem under all these resource constraints. Our approach fills this gap. There are very few studies on materialized view selection in a distributed context. The most efficient work on this field is [7]. They proposed a practical approach that solves the view selection problem with a genetic algorithm, a type of randomized algorithms. The study described in [7] seems to be the only one which addresses the problem in large and complex scenarios that represent real world problems. Genetic algorithms can be applicable on the large search space. They can find a reasonable solution within a relatively short period of time by trading execution time for quality. However, there is no guarantee of performance because of their probabilistic behavior. Besides, the quality of the solution depends on the set-up of the algorithm as well as the extremely difficult fine-tuning of the algorithm that must be performed during many test runs. To provide better query-processing efficiency with respect to the currently most efficient approach [7], we have proposed a novel approach that is based on constraint programming techniques.

Constraint programming is known to be efficient for the resolution of NP-complete problems and a powerful method for modeling and solving combinatorial optimization problems [36]. We have demonstrated in our previous work [30,31] the benefit of using constraint programming techniques for solving the view selection problem with reference to the centralized context. We have also exploited the capabilities offered by the constraint programming paradigm for the benefit of the distributed management for data analysis [29]. In this primary work, we have modeled the view selection problem in a distributed context as a constraint optimization problem in order to solve it by means of constraint programming. We have also been able to easily handle and model all the resource constraints with the rich constraint programming language. Our first results confirm the efficiency of our approach for recommending a high quality set of materialized views. However, our approach grows out of memory and fails to produce a solution within a reasonable execution time for large scenarios.

The reason is that we have considered the complete solution space of all possible view combinations. Obviously, the complexity of a complete search is extremely high: the number of possible views to materialize grows exponentially with the number of sites and queries as well as with the number of columns, join predicates, grouping clauses and the base relations referenced in each query. Consequently, we have evaluated our approach with only simple scenarios consisting of small number of sites (i.e., 20 sites). Due to the huge solution space, constraint solvers cannot be applied directly to solve the view selection problem in complex and large scenarios. This would lead to the need of robust and efficient heuristic search strategies which may reduce significantly the solution space so that the problem becomes solvable.

Goals and Contributions. Our aim is to improve the processing efficiency of queries in distributed relational data-intensive systems by using materialized view techniques. The goal is to propose the most appropriate views to be materialized at the right sites, in order to optimize the combined cost of query evaluation, view maintenance, view storage, network reads and writes, CPU utilization and I/O.

We have formulated and solved this problem, drawing connections to constraint satisfaction and optimization problems studied in both artificial intelligence and operations research. The focus of this paper is to improve the materialized-views search strategy and confirm that our approach scales very well with very large distributed environments.

Our key contributions include designing and developing efficient heuristics to reduce the solution space of candidate views to materialization. We start by considering only efficient query plans and discarding those which are very costly for executing the query workload in the most cost-effective way. To do this, we have been inspired by the *join ordering* heuristic that is an important aspect of centralized query optimization [40]. *join ordering* in a distributed context is even more important since joins between fragments may increase the communication time [34] which is considered to be the dominant parameter in a distributed context. The query processor must also select the best sites to process data, and where the data should be materialized in order to avoid having to pay an important cost of communication. For this reason, we have designed *site selection* heuristics which select the most promising sites on which the views may be processed or materialized. We have also defined heuristic search strategies within the constraint solver to restrict its search space. This could lead to guide the search close to the optimal solution and find near-optimal solutions in a small amount of time. The most common branching strategies in constraint solvers are *variable ordering heuristics* and *value ordering heuristics*. We have designed these heuristics in the way that constraint solver starts by placing a view closest to where it is frequently accessed, while considering resource constraints.

We have proved the effectiveness of these heuristics and their improvements over our approach to trade off completeness for efficiency of the search. The experiment results show the benefit of using these heuristics to reduce the solution space, small enough so that the view selection problem becomes solvable in large-scale scenarios. We have evaluated and compared our approach with a practical solution that uses a genetic algorithm to solve the view selection problem in a large-scale distributed environment [7]. We experimentally show that our approach is able to achieve significant performance gains in comparison with the genetic algorithm. The results of these experiments have also shown the robustness and scalability of our approach. We are able to propose a high quality set of materialized views for very large scenarios consisting of big relations containing millions of tuples and distributed over a large number of sites, which issue large query workloads.

Outline. The rest of this paper is organized as follows. Section 2 reviews related work. In Sect. 3, we formally define the view selection problem in a distributed context and discuss the settings for the problem. Section 4 provides an overview of constraint programming techniques and describes how to model the view selection problem as a constraint optimization problem. Section 5 presents the heuristic search strategies that we have designed for optimization purpose. In Sect. 6, we provide our experimental evaluation. Section 7 contains concluding remarks and future research directions.

2   Related Work

Query performance optimization based on views has been intensely studied for a number of years. One line of past research considers the view selection in a centralized environment. In such scenarios, heuristic algorithms to the well studied problem of view selection can be classified into three main categories. The first category of deterministic view selection algorithms takes a deterministic approach either by exhaustive search or by some heuristics such as greedy [5,13,14,16,24,26,41,42]. These techniques have been shown to consider a maximum storage overhead and/or maximum maintenance costs when deciding which views to materialize.

The second main category of randomized view selection algorithms either uses genetic algorithms [17,23,25,44,45] or simulated annealing [8,9,18] to solve the view selection problem. Genetic algorithms are inspired by the natural evolution process such as selection, mutation, and crossover. The search strategy for these algorithms is very similar to biological evolution while simulated annealing algorithms are motivated by an analogy to annealing in solids. In contrast with simulated annealing algorithms, genetic algorithms use a multi-directional search which allows to efficiently search the space and find better solution quality [25]. The success of randomized algorithms often depends on the set-up of the algorithm as well as the extremely difficult fine-tuning of the algorithm that must be performed during many test runs. To overcome this, we designed a solution that is based on constraint logic programming in which the user has only to specify the problem itself instead of specifying how to solve the problem.

The last category focuses on hybrid view selection algorithms. These algorithms combine the strategies of deterministic and randomized algorithms in their search in order to provide better query performance. A hybrid approach has been applied in [46] to solve the view selection problem. Their experimental results confirmed that hybrid algorithms provide better performance in terms of solution quality. However, their algorithms are more time consuming and may be impractical due to their excessive computation time.

The view selection algorithms, which have been proposed so far to facilitate efficient query processing, may provide near-optimal solutions but there is no performance guarantee because of their greedy nature or their probabilistic behavior. We innovate by proposing an approach that can guarantee to have a set of recommended views to be materialized at any time and seek the optimal solution while the computational time is not restricted.

Past work has also focused on dynamic view selection [22,47]. To respond to the changes in the query workload over time, views have to be selected continuously and replaced with more beneficial views. However, the task of monitoring constantly the query pattern and periodically recalibrating the views is rather complicated and time consuming especially in large data warehouse where many users with different profiles submit their queries. This is different with our study since we consider a given workload where queries are assumed to be known and with each query it is associated a frequency of occurrence. A variation of this setting are caching approaches [10,38]. With caching, the cache is initially empty and data are inserted or deleted from it during the query processing. Materialization could be performed even if no queries have been processed and materialized views have to be updated in response of changes on the base relations. A detailed comparison of these two techniques is given in [20].

Improving query performance is also being studied in distributed databases and data warehouses. Returning to view caching setting, some research works have adopted caching in distributed databases [21] and peer-to-peer systems [19]. Dynamic materialized view selection has also been explored in peer-to-peer environment [6]. However, the results of this study were not validated through an experimental evaluation. Investigating the view selection problem in a dynamic and distributed system is what we are planning to do as future work which is one of challenging research directions. The studies in [3,7,12,43] focus on the problem of materializing the right views at the appropriate sites while considering static queries. The authors in [12] deal with the problem in peer-to-peer environment. In fact, it is provided a full definition of the problem but without providing any algorithm or detail on how to select an effective set of views to materialize and place them at appropriate peers. The works published in [3,43] address the view selection problem in a distributed data warehouse environment. They extend a greedy-based selection algorithm for the distributed case. However, the cost model that they have used does not include the network transmission cost which is very important in a distributed context. The study presented in [7] deals with the view selection problem in distributed databases. This approach consists in applying a genetic algorithm to compute which view has to be materialized at what site of the network. However, this approach does not take into account the resource constrains. In contrary, our approach considers the storage space capacity associated to each site of the network as well as the view maintenance time while making view selection and data placement decisions. The results of the work in [7] are scalable to large-scale distributed scenarios. For these reasons, we have extended our approach, on the results of [29], by designing a set of pruning heuristics which may be used to prove the scalability of our approach in distributed scenarios.

For a deeper overview of view selection methods in a centralized environment as well as in a distributed scenario, we may refer the reader to the survey that we have done in our previous work [28].

3   Formal Setting and Problems

3.1   View Selection in a Distributed Context

We define the view selection for a distributed system as follows. Assume we have given a set of cooperating sites connected by a network with different resource constraints on CPU, IO and network bandwidth. Sites issue numerous queries and exchange data with a collection of participating sites. To minimize the total evaluation cost of the queries, we may materialize a set of views at a collection of sites and use these views to answer the queries. We assume that we have a storage space limit at each site that is the total space of views must not exceed the one associated to the site where they should be materialized. The view selection is also constrained by a maintenance cost limit to keep the materialized views in synchronization with the underlying base data. The general problem of view selection in a distributed context can be formally formulated as follows.

Definition 3.1 Distributed View Selection Problem (DVSP). *Let $S = \{s_1, s_2, ..., s_s\}$ be a set of sites in the network where each site s_i has associated IO and CPU resources and storage space Sp_{max_i}. Every pair of sites s_i and s_j is connected by a bandwidth-constrained link. Let $Q = \{q_1, q_2, ..., q_q\}$ the query workload we expect to be given and $fQ = \{f_{q_1}, f_{q_2}, ..., f_{q_q}\}$ their query frequency. Each query has an associated non-negative weight which describes its frequency relative to the combined workload across all sites. Assume that $U = \{u_1, u_2, ..., u_u\}$ is the set of updates on base data, $fU = \{fu_{r_1}, fu_{r_2}, ..., fu_{r_r}\}$ their update frequency and U_{max} the maintenance time limit. The DVSP consists in finding candidate view set $\langle V, S \rangle$ such that the cost of evaluating Q is minimal.*

An overview of the architecture of our system is shown in Fig. 1. Sites are connected by bandwidth-constrained links and cooperate to materialize views and answer queries. In our approach, sites may serve any or all of the following four roles:

- *Data origin* S_{DO} provides original content (base relations) and is the authoritative source of that data.
- *Storage provider* S_{SP} stores materialized views if there is enough storage space.
- *Query evaluator* S_{QE} uses a portion of its CPU and IO resources to evaluate the set of queries forming its workload.
- *Query initiator* S_{QI} acts as a user in the system and issues queries.

To illustrate, we present the following example for a network with five sites.

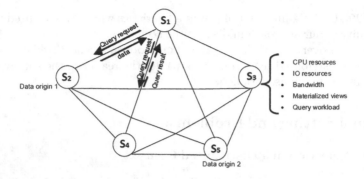

Fig. 1. Overview of our system architecture: Simple example scenario

Example 3.1 *Given a distributed database scenario consisting of five sites s_1, s_2, s_3, s_4 and s_5 which are connected by a network that has limited bandwidth (see Fig. 1). Recall from Sect. 1 that the query q_1 is posed frequently at site s_1, and we have used the view v_1 which is materialized at site s_1 to speed up the evaluation of query q_1. Suppose that the view v_1 is too large to fit in the available storage space at the site s_1 or the base relations referenced in query q_1 are frequently updated and the cost that we save for answering the query is now involved for the maintenance cost. In this case, materializing v_1 at site s_1 cannot be attractive because of the space or maintenance cost constraint. The cooperative decisions between sites should also be taken into account during the view selection and data placement process. Assume in our example that the connection speed between (s_1, s_4) and (s_2, s_4) is very high and the resources in site s_4 are under-utilized while the connection speed between (s_1, s_3) and (s_1, s_5) is very low. In this case, it is not interesting to materialize v_1 at s_3 or s_5 but it may be beneficial to materialize v_1 at s_4. Hence, to materialize the right set of views (or queries) at the appropriate sites, the view selection problem has to be studied under the existing resource and bandwidth constraints.*

3.2   The Space of Candidate Views

For modeling the space of possible candidate views, our approach use the distributed AND-OR view graph which is an extension of the concept of the AND-OR view graph to capture the distributed case. The AND-OR view graph framework has been used as a tool for setting up the search space by identifying common sub-expressions between the different queries of the workload [13–15, 32, 37] (considering the possibility of reusing parts of query plans). Recognize possibilities of shared views can be exploited for sharing computation, updates and storage space. Hence, exploiting common sub-expressions can help speed up query and update processing. We borrow the rules provided in [37] for identifying common sub-expression. In what follows, we start by giving a definition of the AND-OR view graph.

Definition 3.2 (AND-OR View Graph). *Given a query set Q defined over a database schema R, the AND-OR view graph $G = (N, E)$ is a DAG, where N represents nodes and E represents edges, such that:*

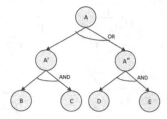

Fig. 2. AND-OR DAG representation

- $N = N_{Eq} \cup N_{Op}$. The node type of G is an operation node (Op-Node) or an equivalence (Eq-node). Each Op-node $op_i \in N_{Op}$ represents an algebraic expression (Select-Project-Join) with possible aggregate function. An Eq-Node $eq_i \in N_{Eq}$ represents a set of logical expressions that yield the same result. In other words, equivalence nodes correspond to each view that is candidate to the view selection. The leaf nodes are equivalence nodes representing the base relations R and the root nodes represents the result of every query q in Q.
- The dependence over the nodes are represented by OR-Arc or AND-Arc through the edges. The first dependence means that a parent node can be computed for any one of its children nodes. For example, in Fig. 2, node A can be computed from any of the children nodes A' and A''. While, the later dependence specifies that the children nodes are required to compute the parent node. For instance, in Fig. 2, Nodes B and C are needed to compute A'. For each query q in Q, there is an AND-OR-DAG representation which consists of its all possible execution plans.

The AND-OR view graph of two queries, are shown in Fig. 3. In addition to q_1 presented previously in Sect. 1, we consider another query q_2 which finds the number of products having as a brand name 'BMW' and bought by the united states.

q_2:**Select P.name, count(*)**
   From PartSupp PS, Supplier S, Nation N, Part P
   Where PS.suppkey=S.suppkey and S.nationkey=N.nationkey and
   P.partkey=PS.partkey and N.name='USA' and P.brand='BMW'
   Group By P.name;

For simplicity, we consider in our example (see Fig. 3) only one way to answer the query q_2. The remaining execution plans are just indicated by dashed lines. Two ways for evaluating q_1 have been considered by applying the join commutativity and associativity rules. For more details about constructing the AND-OR view graph for the query workload, we may refer the reader to [37]. Circles represent operations nodes (Op-Nodes) and boxes represent equivalence nodes (Eq-Nodes). The OR-Arc and AND-Arc are indicated by drawing a semicircle, through the edges. The subscripts PS, S, N and P denote respectively the base relations of TPC-H benchmark: PartSupp, Supplier, Nation and Part.

In the AND-OR view graph, the choice of materialized views is done in conjunction with choosing execution plans for queries which is very important to optimize

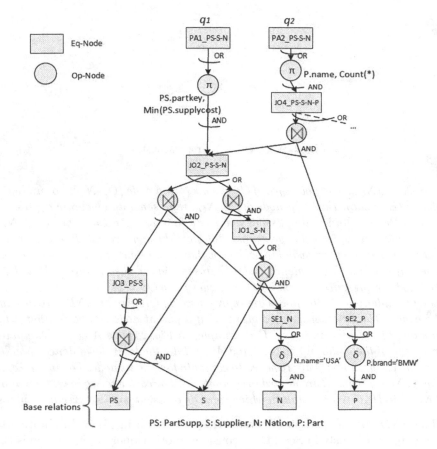

Fig. 3. AND-OR view graph of two queries q_1 and q_2

query performance. For instance, a plan that seems quite inefficient could become the best plan if some intermediate result of the plan is chosen to be materialized. To deal with distributed settings and reflect the relation between views and communication network, we propose the distributed AND-OR view graph, which can be seen as an extension of the concept of the AND-OR view graph. The distributed AND-OR view graph can be defined as follows.

Definition 3.3 (Distributed AND-OR View Graph). *Let $G = (N', E')$ the distributed AND-OR view graph so that:*

- $N' = N_{(Eq,S)} \cup N_{(Op,S)}$ *where each node $(eq_i, s_j) \in N_{(Eq,S)}$ denotes a potential site at which a view may be materialized and $(op_i, s_j) \in N_{(Op,S)}$ indicate the decision on what site the algebraic operation can be evaluated. To include this representation, every node of the AND-OR view graph is split into sub-nodes, each of which denotes the evaluation at a given site.*
- $N_{(Eq,S)} = N^o_{(Eq,S)} \cup N^t_{(Eq,S)}$.*Further edges are introduced in E' which are the communication edges between equivalence nodes of the same level, denote that a*

view can be answered from any other site if it is less expensive. However, cycles occur in the resulting distributed AND-OR view graph (which no longer conforms to the characteristics of a DAG). To eliminate cycles, each equivalence node (eq_i, s_j) has to be split into two nodes $(eq_i, s_j)^o$ (origin node) and $(eq_i, s_j)^t$ (target node).

A sample distributed AND-OR view graph is shown in Fig. 4. For simplicity, we consider a network of only five sites s_1, s_2, s_3, s_4 and s_5 depicted in Fig. 1 and we illustrate a part of the query q_1 by considering only one execution strategy $(\Pi((\delta_N \bowtie S) \bowtie PS))$. The communication edges between equivalence nodes of the same level (i.e., $(PA1\_PS\text{-}S\text{-}N, S_2)$), $((PA1\_PS\text{-}S\text{-}N, S_3))$ and $((PA1\_PS\text{-}S\text{-}N, S_4))$), are shown in the dashed rectangle in Fig. 4. However, these edges are bidirectional creating cycles. To eliminate these cycles, each equivalence node (v_i, s_j) in the distributed AND-OR view graph, as illustrated in Fig. 5, has been artificially split into two nodes $(v_i, s_j)^o$ and $(v_i, s_j)^t$.

3.3   Estimated View Cost

To each query or view (v_i, s_j) in the distributed AND-OR view graph, we associate a cost estimation about the query cost that is the evaluation cost of the cheapest execution strategy for (v_i, s_j), the maintenance cost which is the cost of the efficient plan for maintaining the view (v_i, s_j)) and the reading cost that is the cost of reading v_i on site s_j. In a distributed system, a cost model should reflect CPU, I/O and communication costs. In our approach, the costs are estimated in terms of time. The main factor for estimating the different costs is to estimate the size of any possible view created throughout the search.

View Size. To estimate the size of a given view $(v_i, s_j) \in \langle V, S \rangle$ we adopt the solution of [34], which is based on database statistic and known relational formulas to predict the cardinalities of the results of relational algebra operations. Two simplifying assumptions are commonly made about the database: (i) the distribution of attribute values in a relation is supposed to be uniform, and (ii) all attributes are independent, meaning that the value of an attribute does not affect the value of any other attribute. We define the size of the view v_i as follows.

$$size(v_i, s_j) = size(v_i) = card(v_i) * length(v_i)$$

where $length(v_i)$ is the length (in number of bytes) of a tuple of v_i, computed from the lengths of its attributes and $card(v_i)$ is the number of tuples in v_i.

Reading Cost. This cost is considered if a given view has been selected to be materialized on a given site. The cost of reading a view v_i on site s_j is estimated as follows.

$$RCost(v_i, s_j) = size(v_i) * I/O_j$$

Communication Cost. In a distributed system, the communication cost is considered to be the dominant factor. It reflects the time needed for exchanging data between sites participating in the execution of the query. For instance, given a

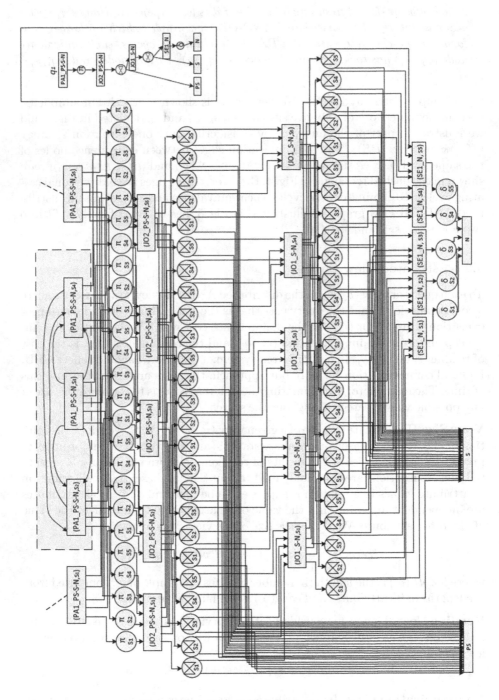

Fig. 4. Distributed AND-OR view graph

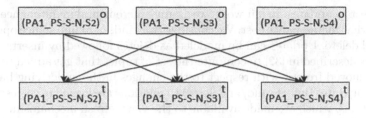

Fig. 5. Applying split to eliminate cycles

query q_i which is issued at the site s_j and denoting by v_k, a view required to answer q_i, the communication cost is zero if v_k is materialized at s_j. Otherwise, let s_l be the node containing v_k, then the communication cost for transferring v_k from s_l to s_j is:

$$CCost(v_k, s_l \rightarrow s_j) = \frac{size(v_k)}{Bw(s_j, s_l)}$$

where $Bw(s_j, s_l)$ is the bandwidth between s_j and s_l (i.e., network transmission cost per unit of data transferred) and $size(v_k)$ is the size of v_k in number of bytes.

Query Cost. This cost estimation refers to the amount of time necessary to compute the answer to a given query or view using the most efficient execution strategy. It includes CPU, I/O, and communication costs. The CPU cost is estimated as the time needed to process each tuple of the relation e.g., checking selection conditions. The IO cost estimate is the time necessary for fetching each tuple of the relation. For the evaluation cost of relational operations, we use a cost model similar to [7, 11, 27, 39]. Note that for each relational operator, we consider its most simple implementation, e.g., sequential scans and nested loop joins.

  Estimated Cost of Relational Operations.

– Estimated Cost of Unary Operations
  - $cost(op_i) = (IO * card * length) + (CPU * card * lengthP)$ where op_i is a selection operation
  - $cost(op_i) = (IO * card * log(card) * length) + (CPU * card * log(card) * lengthP)$ where op_i is a projection operation
  - $cost(op_i) = (IO * card * length) + (CPU * card * lengthA)$ where op_i is an aggregation operation
– Estimated Cost of Binary Operations
  - $cost(op_i) = (IO * lcard * rcard * (llength + rlength)) + (CPU * lcard * rcard * lengthP)$ where op_i is a join operation

Where $card$ is the number of tuples of the operand, $length$ is the length (in bytes) of a tuple, $lengthP$ is the length of columns checked by predicates, $lengthA$ is the length of the tuples being aggregated, $lcard$ and $rcard$ are respectively the number of tuples of the left and right operands (the same for $llength$ and $rlength$).

View Maintenance Cost. It is the cost required for updating views when the related base relations are changed. The view maintenance cost is computed similarly to the query cost, but the cost of relational operation evaluation is computed with respect to updates. In our work, we assume incremental maintenance to estimate the view maintenance cost. We consider two kinds of maintenance operation: insert and delete. Updates can be modeled as deletes followed by inserts. We use techniques described in [32] to compute the set of tuples that get added to a given view or removed from it with respect to the changes in the underlying base relations. The updates (inserts and/or deletes) to relations are logged in corresponding *delta* relations, which are made available to the view refresh mechanism.

Section 4.1 details how we compute the query cost $Qc(v_i, s_j)$ and maintenance cost $Mc(v_i, s_j)$ corresponding to each view (v_i, s_j) (or query) in the distributed AND-OR view graph.

4   Constraint Programming Method for the DVSP

Constraint programming, originated from Artificial Intelligence, is a solution method to the combinatorial optimization problems. It has been considered as beneficial in data mining setting [35] and successfully applied in numerous combinatorial search problems [36] such as scheduling and timetabling. Constraint Programming has the strength of modeling the problem by stating constraints in a declarative way, which must be satisfied by the solution, without the need of being interested in the way the problem is solved. By constraint programming, we mean the techniques that are used to represent and solve the Constraint Satisfaction Problem (CSP) and Constraint Optimization Problem (COP) arising from Artificial Intelligence.

4.1   Background

We now give the basic notation of constraint programming.

Definition 4.1 (Variable and Domain). *Let $VAR = \{var_1, var_2, ..., var_n\}$ be the set of variables of the problem. The domain $DOM = \{d_{var_1}, d_{var_2}, ..., d_{var_n}\}$ is the set of possible values that can be assigned to each variable var_i. A single value is assigned to a variable*

Example 4.1 $Mat(v_1, s_1)$ *and* $Mat(v_2, s_1)$ *are variables and their respective domains are* $d_{Mat(v_1, s_1)} = d_{Mat(v_1, s_2)} = \{0, 1\}$. $Mat(v_i, s_j)$ *denotes for each view v_i if it is materialized or not materialized on site s_j (0: v_i is not materialized at site s_j, 1: v_i is materialized at s_j)*

Definition 4.2 (Constraint). *Consider a finite sequence of variables $VAR = \{var_1, var_2, ..., var_n\}$, with respective domains $DOM = \{d_{var_1}, d_{var_2}, ..., d_{var_n}\}$. A constraint C_{ijk} between the variables var_i, var_j, var_k is any subset of the possible combinations of values of var_i, var_j, var_k, i.e., $C_{ijk} \subset d_{var_i} \times d_{var_j} \times d_{var_k}$*

Example 4.2 *On variables $Mat(v_1, s_1)$ and $Mat(v_2, s_1)$, we impose the binary constraint $Mat(v_1, s_1) * size(v_1) + Mat(v_2, s_1) * size(v_2) \leq 5$. The constraint states that the size of materialized views v_1 and v_2 at site s_1 should be less or equal that 5*

Definition 4.3 (Constraint Satisfaction Problem (CSP). *A CSP, is defined by a finite sequence of variables $VAR = \{var_1, var_2, ..., var_n\}$ with respective domains $DOM = \{d_{var_1}, d_{var_2}, ..., d_{var_n}\}$, together with a finite set of constraints $CST = \{c_1, c_2, ..., c_n\}$, each on a subsequence of VAR. Therefore, a CSP is defined by a triplet (VAR;DOM;CST). A feasible solution to a CSP is an assignment of a value from its domain to every variable, so that the constraints on these variables are satisfied.*

Example 4.3 *Based on the variables given in Example 4.1 and constraints given in Example 4.2, we denote the resulting CSP as:*

$$Mat(v_1, s_1) \in \{0,1\}, Mat(v_2, s_1) \in \{0,1\}$$

$$Mat(v_1, s_1) * size(v_1) + Mat(v_2, s_1) * size(v_2, s_1) \leq 5$$

Definition 4.4 (Constraint Optimization Problem (COP)). *A COP is a CSP with an objective function $f : d_{var_1} \times d_{var_2} \times d_{var_n} \to R$ to be optimized. An optimal solution to a COP is a solution to CSP that is optimal with respect to f. The cost expression on these variables takes a maximal or minimal value for maximization or a minimization problem, respectively*

In our work, the DVSP is modeled as a COP in order to solve it by means of constraint programming. The goal is to decide which view has to be materialized at what site to optimize the query processing.

The solution process in constraint programming interleaves *constraint propagation*, and *search*. Actually, this is the main strength of constraint programming. The solution approach to combinatorial optimization problems requires the exploration of the search space represented by all the possible combinations of the assignments of values to the variables. However, some parts of the search space can be pruned (no need to be visited). This can be done by efficient propagation algorithms associated with constraints. *Constraint propagation* (also called the feasibility prune) is based on the feasibility reasoning which removes the assignments of values to variables that do not lead to any feasible solutions [4,36]. When the *search* fixes the value of a variable, constraint propagation is applied to restrict the domains of other variables whose values are not currently fixed. This means that when a value is assigned to the current variable, any value in the domain of a future variable which conflicts with this assignment is removed from the domain. In constraint programming, *constraint propagation* and *search* are applied in an alternated fashion.

Let us now illustrate how constraint programming can be applied to select and place the right views at the appropriate sites (see Fig. 6). At the beginning, the initial variable domains $d_{Mat(v_i, s_j)} = \{0,1\}$ where $i, j \in \{1, 2, 3\}$. Recall that $Mat(v_i, s_j)$ denotes for each view v_i whether it is materialized or not materialized at site s_j (0: v_i is not materialized at site s_j, 1: v_i is materialized at s_j). The problem is to select a set of views and a set of sites at which these views should be materialized subject to a space and maintenance constrains. The space constraint ensures that the total space occupied by the materialized views at each site

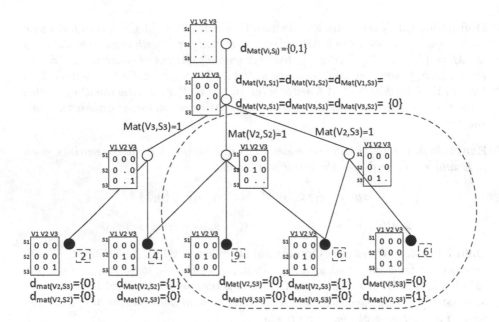

Fig. 6. Using constraint programming to solve the DVSP: *constraint propagation* and search

is less or equal than its storage space capacity $(\forall s_j \in S \sum_{(v_i,s_j)\in V}(Mat(v_i,s_j) *$ $size(v_i)) \leq Sp_{max_{s_j}})$. Let as assume that $Sp_{max_{s_1}} = 4\,\text{MB}$, $Sp_{max_{s_2}} = 10\,\text{MB}$, $Sp_{max_{s_3}} = 15\,\text{MB}$, $size(v_1) = 10\,\text{MB}$, $size(v_2) = 5\,\text{MB}$ and $size(v_3) = 14\,\text{MB}$. Considering the space constraint, it appears for example that the view v_1 cannot be materialized at site s_1 since $size(v_1) \succ Sp_{max_{s_1}}$. The maintenance cost constraint guarantees that the total maintenance cost of the set of materialized views is less or equal than the maximum view maintenance cost U_{max} $(\sum_{(v_i,s_j)\in V}(Mat(v_i,s_j) *$ $f_u(v_i) * Mc(v_i,s_j)) \leq U_{max})$. Assuming that $U_{max} = 20\,\text{s}$, $Mc(v_1,s_1) = 13\,\text{s}$, $Mc(v_1,s_2) = 24\,\text{s}$, $Mc(v_1,s_3) = 21\,\text{sec}$, $Mc(v_2,s_1) = 3\,\text{s}$, $Mc(v_2,s_2) = 8\,\text{s}$, $Mc(v_2,s_3) = 5\,\text{s}$, $Mc(v_3,s_1) = 6\,\text{s}$, $Mc(v_3,s_2) = 15\,\text{s}$ and $Mc(v_3,s_3) = 10\,\text{s}$; where $Mc(v_i,s_j)$ denotes the cost of maintaining the view v_i on site s_j. The update frequencies $f_u(v_i)$ are at scale 1. It appears that $Mat(v_1,s_2)$ and $Mat(v_1,s_3)$ cannot take the value 1 because otherwise the total maintenance cost will be greater than U_{max}. This value is eliminated from the variable domain by applying *constraint propagation*. In a similar fashion the inconsistent value 1 is removed from the domain of $Mat(v_1,s_1)$, $Mat(v_2,s_1)$, $Mat(v_3,s_1)$ and $Mat(v_3,s_2)$ by taking into account the space constraint.

After this stage the variable domains are not reduced to singletons (there is more than one value in each variable's domain), the solver takes during the *search* step one of these variables and tries to assign it each of the possible values in turn e.g., $Mat(v_3,s_3) = 1$. This enumeration stage triggers more variable domain reductions by using constraint propagation techniques which leads in our example to five solutions that are marked with a black circle. These solutions are of various cost or quality. The costs are indicated in the small dashed rectangle.

In addition to providing a rich constraint language to model a problem as a COP and techniques such as *search* and *constraint propagation* to reduce the search space by excluding solutions where the constraints become inconsistent, constraint programming offers facilities to control the search behavior. This means that search strategies can be defined to decide in which order to explore the created children nodes in an enumeration tree. This avoid visiting a large subset of unpromising candidate solutions during the tree search. Returning to our example illustrated in Fig. 6, a well-suited search strategy would reduce the number of expanded nodes in the tree search by avoid visiting the solutions that are in the dashed block. In Sect. 5, we present our own search strategy that we have designed within the constraint solver to speed up the search to near-optimal or optimal solutions for the DVSP. Constraint programming also provides ways to limit the tree search regarding different criteria. For instance performing the search until reaching a feasible solution in which all constraints are satisfied, or until reaching a search time limit or until reaching the optimal solution.

4.2 Modeling the DVSP as a COP

In this subsection, we describe how to model the view selection problem in a distributed scenario as a Constraint Optimization Problem (COP). Then, its resolution is supported automatically by the constraint solver embedded in the constraint programming language. In what follows, we define all the symbols as well as the variables that we have used in our COP. A prior version of this COP was demonstrated in [29].

- G. The distributed AND-OR view graph for the queries of workload.
- $Q(G)$. The query workload which models our expected queries.
- $V(G)$. The set of candidate views that is highly dependent on the set of queries we expect to be given.
- U. The set of updates (inserts/deletes) to base relations.
- $\delta(v_i, s_j, u)$. The differential result of view v_i on s_j (i.e., the set of tuples inserted into and/or deleted from (v_i, s_j)), with respect to update u.
- f_q. The non-negative weight associated to query q which describes its relative frequency within the workload.
- f_u. The update frequency that indicates the frequency of updating a given view.
- S. The set of sites which represents the computer nodes over the network.
- $Sp_{max_{s_i}}$. The storage space capacity of the site s_i.
- U_{max}. The maximum view maintenance cost which is the time window available to provide up-to-date responses to queries.
- $size(v_i)$. The size of the view v_i in terms of number of bytes.
- $Bw(s_k, s_j)$. The bandwidth between s_j and s_k that represents the network transmission cost per unit of data transferred.
- $Mat(v_i, s_j)$. The materialization variable that denotes for each view (v_i, s_j) (equivalence node in the distributed AND-OR view graph G), if it is materialized or not materialized. It is a binary variable, $d_{Mat_{(v_i, s_j)}} = \{0,1\}$.

- $Qc(v_i, s_j)$. The query cost corresponding to the view v_i if it is evaluated or materialized on site s_j. The domain is a finite subset of $\mathbb{R}_+^*$ such as $d_{Qc(v_i,s_j)} \subset \mathbb{R}_+^*$
- $Mc(v_i, s_j)$. The maintenance cost corresponding to the view v_i if it is updated on site s_j, where $d_{Mc(v_i,s_j)} \subset \mathbb{R}_+^*)$.

The view selection in a distributed scenario can be formulated by the following constraint optimization model.

$$minimize \sum_{(v_i,s_j)\in Q(G)} \left(f_q(v_i) * Qc(v_i, s_j) \right) \tag{1}$$

$$subject\ to \quad \forall s_j \in S \sum_{(v_i,s_j)\in V(G)} \left(Mat(v_i, s_j) * size(v_i) \right) \leq Sp_{max_{s_j}} \tag{2}$$

$$\sum_{(v_i,s_j)\in V(G)} \left(Mat(v_i, s_j) * f_u(v_i) * Mc(v_i, s_j) \right) \leq U_{max} \tag{3}$$

In our approach, the main objective is the minimization of the weighted query processing cost. The total query cost is computed by summing over the cost of processing each input query rewritten over the materialized views. Constraints (2) and (3) state that the distributed view selection problem is studied given a limited amount of resources e.g., storage space and maintenance time. Constraint (2) ensures that for each site the total space with respect to materialized views on it is at most equal to its storage space capacity. Constraint (3) guarantees that views are selected to be materialized under the constraint that their update costs is less or equal than the total view maintenance cost.

The query and maintenance costs corresponding to a view are implemented by using a depth-first traversal of the distributed AND-OR view graph. Note that the query and maintenance costs corresponding to base relations (leaf nodes) are equal to zero. They may be formulated as follows.

$$Qc(v_i, s_j) = \min_{s_k \in S} \left(Qc_{local}(v_i, s_k) + \frac{size(v_i)}{Bw(s_k, s_j)} \right) \tag{4}$$

$$Qc_{local}(v_i, s_j) = \begin{cases} ComputeCost(v_i, s_j) & if\ Mat(v_i, s_j) = 0 \\ size(v_i) * I/O_{s_j} & otherwise \end{cases} \tag{5}$$

$$ComputeCost(v_i, s_j) = \min_{op_l \in child(v_i,s_j)} \left(cost(op_l, s_j) + \right.$$

$$\left. \sum_{(v_m,s_n)\in child(op_l)} \left(Qc(v_m, s_n) + \frac{size(v_m)}{Bw(s_n, s_j)} \right) \right) \tag{6}$$

Query Cost. The query cost corresponding to each given view in the AND-OR view graph in a distributed system reflects the local processing cost and the communication cost. The first component is estimated with respect to the evaluation costs of the required relational operations (see Sect. 3.3). The communication cost is computed as the time needed for transmitting views on the communication network. Constraint (4) guarantees that sites which can provide responses with lowest cost are selected to answer views (or queries). Constraint (5) and (6) ensure that the most efficient execution plans are chosen to compute query results. Each query plan is composed of all the cost of executing the relational operations (operation nodes in the distributed AND-OR view graph) on the path. The reading cost is considered if the view has been selected to be materialized.

$$
Mc(v_i, s_j) = \begin{cases} 0 \ if \ Mat(v_i, s_j) = 0 \\ \sum_{u \in U} \left(\min_{s_k \in S} \left(Mcost(v_i, s_k, u) + \frac{size(v_i)}{Bw(s_k, s_j)} \right) \right) otherwise \end{cases}
$$

(7)

$$
Mcost(v_i, s_j, u) = \min_{op_l \in child(v_i, s_j)} \left(cost(op_l, s_j, u) + \right.
$$

(8)

$$
\left. \sum_{(v_m, s_n) \in child(op_l)} \left(UpdateCost(v_m, s_n, u) + \frac{size(v_m)}{Bw(s_n, s_j)} \right) \right)
$$

$$
UpdateCost(v_m, s_n, u) = \begin{cases} Mcost(v_m, s_n, u) + \frac{size(v_l)}{Bw(s_n, s_m)} \ if \ Mat(v_m, s_n) = 0 \\ \delta(v_m, s_n, u) * I/O_{s_n} \ otherwise \end{cases}
$$

(9)

View Maintenance Cost. Because materialized views have to be kept up to date, the view maintenance cost has to be considered during view selection and data placement process. The maintenance cost is the differential results of materialized views given the updates of the bases relations. Constraint (7) guarantees that there is no maintenance cost if the view has not been materialized. Otherwise, this cost is computed by summing the number of changes in the base relations from which the view is updated. A view (or a query) is updated from the site that can provide the differential results with the lowest cost. As mentioned in Sect. 3.3, we assume incremental maintenance to estimate the view maintenance cost. Constraints (8) and (9) insure that the best plan with the minimum maintenance cost is selected in order to optimize the update of a set of materialized views. The view maintenance cost is computed similarly to the query cost, but the cost of each update plan is composed of all the cost of executing the relational operations with respect to updates corresponding to the related base relations. The maintenance cost is considered equal to the cost of reading the changes over the view if the latter has been materialized.

5 Heuristics to Reduce the Search Space of Views

In this Section, we discuss search strategies for navigating in the solution space of candidate view sets, looking for a low or minimal query cost subject to resource constraints. The distributed AND-OR view graph defined in Sect. 3.2 which is one of the inputs to the DVSP, represent all possible execution strategies for each query in the workload. Clearly, the solution space is huge by considering all possible execution query plans. Indeed, the number of possible views to materialize grows exponentially with the number of sites and queries as well as with the number of join predicate and relations referenced in each query. Even if views are selected off line and thus time is not a big concern, it brings real issues due to memory limitations. To solve the DVSP with low memory needs, special pruning heuristics has to be designed in order to reduce the solution space. We start by considering only efficient execution plans and discarding those which are very costly. As this task is of importance similar to query optimization, we have been inspired by the standard heuristic *join ordering* [34,40] to optimize the ordering of joins in distributed queries. The second reduction of the search space of views is based on a *site selection* heuristic which constructs a set of promising candidate sites at which views should be evaluated or materialized. The third reduction defines robust heuristic search strategies within the constraint solver to determine the traversal of the search tree. The most common branching strategies in the constraint solver are based on the assignment of a selected variable to one or several selected values. The objective of defining our own *variable and value ordering* heuristics is to guide the search close to the optimal solution to the DVSP, which leads the solver to find near-optimal solutions very fast.

5.1 Join Ordering Heuristic (JOH)

An important aspect of query optimization is join ordering [34], since permutations of the joins within the query may lead to improvements of orders of magnitude. Join ordering in a distributed environment is even more important since joins between fragments may increase the communication cost which is considered to be the dominant factor in a distributed context. Indeed, minimizing distributed joins is fundamental to minimize data communication. Let us first concentrate on the simpler problem: consider the query $A \bowtie B$, where A and B are relations stored respectively on site s_1 and s_2. The obvious choice of the join-ordering algorithm is to transfer the smaller relation to the site of the larger one. For instance the order (A, B) (*i.e.*, $size(A) < size(B)$) could use the strategy which sends the relation A to site s_2, while the order (B, A) (*i.e.*, $size(B) < size(A)$) could use the strategy that transmitted B to site s_1.

Let us now consider the query q_1 expressed in relational algebra as follows: $\Pi(\delta_N \bowtie S \bowtie PS)$. This query can be executed in many different ways, some of them are shown in Fig. 4. We define the set of query plans for q_i as $QP_i = \cup_j qp_{ij}$ where $j \leq n$ and $j \geq 1$, and n is the number of possible execution strategies for q_i. We denote by $V_{qp_{ij}}$ the set of views in qp_{ij}. In what follows, we describe two possible strategies qp_{11} and qp_{12} for q_1. Note that $v_i \rightarrow s_j$ stands for "view v_i is transferred to site s_j".

qp_{11}:

s_2 computes $SE1\_N{:}\delta_N$

$SE1\_N \rightarrow s_1$

$S \rightarrow s_1$

s_1 computes $JO1\_S\text{-}N : SE1\_N \bowtie S$

$PS \rightarrow s_1$

s_1 computes $JO2\_PS\text{-}S\text{-}N : JO1\_S\text{-}N \bowtie PS$

s_1 computes $PA1\_PS\text{-}S\text{-}N{:}\ \Pi(JO2\_PS - S - N)$

$V_{qp_{11}} = \{SE1\_N, JO1\_S\text{-}N, JO2\_PS\text{-}S\text{-}N, PA1\_PS\text{-}S\text{-}N\}$

qp_{12}:

$PS \rightarrow s_1$

$S \rightarrow s_1$

s_1 computes $JO3\_PS\text{-}S : PS \bowtie S$

s_2 computes $SE1\_N{:}\delta_N$

$SE1\_N \rightarrow s_1$

s_1 computes $JO2\_PS\text{-}S\text{-}N : JO3\_PS\text{-}S \bowtie SE1\_N$

s_1 computes $PA1\_PS\text{-}S\text{-}N{:}\ \Pi(JO2\_PS\text{-}S\text{-}N)$

$V_{qp_{12}} = \{SE1\_N, JO3\_PS\text{-}S, JO2\_PS\text{-}S\text{-}N, PA1\_PS - S - N\}$

To reduce the search space of candidate views by considering only efficient query plans, we have been inspired by the common used join-ordering algorithm.

Our heuristic works as follows: it starts by considering for each possible execution strategy the join views, denoted by JV $JV_{qp_{11}} = \{(JO1\_S\text{-}N), (JO2\_PS\text{-}S\text{-}N)\}$) and then it computes an estimation of their size (e.g. $size(JO1\_S\text{-}N)$ and $size(JO2\_PS\text{-}S\text{-}N)$). A query plan $qp_{ij} \in QP_i$ is considered as a possible execution strategy for a query $q_i \in Q$ only if the size estimation of join results is smaller than a cost threshold:

$size(v_1) * size(v_2)... * size(v_v) \leq K * size_{max}$ $where$ $v_k \in JV_{qp_{ij}}$

The cost threshold requires two metrics: K is a constant where $K \leq 1$ and $K \succ 0$ (If $K=1$, then all execution plan alternatives are considered for each query in the workload) and $size_{max}$ that is computed as follows:

$\max_{qp_{ij} \in QP_i} \left(size(v_1) * size(v_2)... * size(v_v) \right)$ $where$ $v_k \in JV_{qp_{ij}}$

It follows that qp_{11}, as defined above, is considered as a possible execution strategy for q_1 only if: $size(JO1\_S\text{-}N)* size(JO2\_PS\text{-}S\text{-}N) \preceq K * size_{max}$ where $size_{max}$ is computed as follows while considering only the query plans qp_{11} and qp_{12}.

$\max\left(size(JO1\_S\text{-}N)* size(JO2\_PS\text{-}S\text{-}N), size(JO3\_PS\text{-}S)* size(JO2\_PS\text{-}S\text{-}N) \right)$

5.2 Site Selection Heuristic (SSH)

When distributed query optimization is used, either a single site or several sites may participate to the selection of the strategy to be applied for answering the query. However, the number of possible execution strategies is in fact exponential in the number of sites in the network. In the distributed AND-OR view graph that we have presented in Sect. 3.2, every equivalence node (or view) and operation node is spread over all sites to model all possible execution strategies. A practical solution for the distributed view selection must design heuristics which select the most promising sites on which the views may be computed or materialized in order to avoid having to pay an important cost of communication.

Returning to the Example 3.1 described previously, it appears that it is not beneficial to materialize the query q_1 at site s_3 or site s_5. This leads to the reduction of the search space in the distributed AND-OR view graph by discarding the execution strategies involving the sites s_3 and s_5 instead of considering all sites of the network as shown in Fig. 4. This example illustrates the importance of producing execution plans by carrying out site selection. The sites which are not able to communicate with the query issuer with high bandwidth connections may be discarded in the elaboration of the best execution strategy, when the goal is the efficient usage of the available bandwidth. Intuitively, this strategy may be especially attractive for reducing the search space by considering that some sites in the network are not promising and should not be explored during view selection and data placement process. Site selection heuristics consider the following sites to participate to the selection of the execution strategy for each query of the workload:

- $S_{QI}(q_i)$. Answering the query (or view) at the site that issues this query would improve the efficiency of evaluating the query especially when it is posed frequently. However, it is not always possible when the query issuer has not the required data.
- $S_{DO}(q_i)$. Responses to the queries can be obtained from the data origin if the required data can be found only at the data source. One way to reduce the evaluation cost of queries is to pre-compute and store extra relations at the data origin in order to avoid having to transfer large amounts of data through the network.
- $S_{NE}(q_i)$. The effective utilization of site resources justifies the overhead of searching potential neighbor sites. The request may be sent to the neighbors of the query issuer or data origin. This attempts to exploit under-utilized resources that may exist in some sites and exploring the option of materializing at them the query results.

5.3   Variable and Value Ordering Heuristics (VVO)

A key ingredient of any constraint optimization approach is an efficient search strategy. Indeed, defining well-suited heuristic search strategies within the constraint solver can prune the unpromising nodes in the tree search and hence speed up the search to the optimal solution. Recall from Sect. 4 that the search in a constraint programming approach is organized as an enumeration tree, where each node corresponds to a subspace of the search. A search tree in constraint programming is dynamically built by splitting a problem into smaller subproblems. A subproblem is not further decomposed when the node is either pruned by feasibility

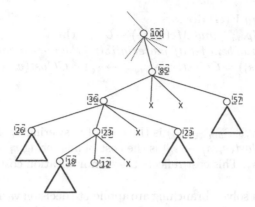

Fig. 7. A search tree for a CSP/COP.

or optimality pruning or a leaf node is reached (as we can see in Fig. 7). The leaf nodes are represented by circles, each of which indicates a possible solution where all variables have value assignments. The costs associated to each solution are indicated in the small dashed rectangle. The feasibility pruning is marked by a cross while the pruned subtree is indicated by a triangle. Indeed, the search subtrees can be pruned where the optimal solution does not settle. Note that every time when a feasible solution is found, such solution imposes an additional constraint, so that further solutions must have a better objective function value.

The tree is progressively constructed by applying series of branching strategies that define the way to branch from a tree search node. The most common branching strategies in the constraint solver are based on *variable and value ordering* heuristics. These heuristics impose an ordering on the variables and values respectively. The order in which variables and values are selected has a great impact on the search process. An example of variable ordering heuristic is the most constrained first variable heuristic. It orders the variables with respect to the number of their appearance in the constraints. An example of value selection ordering is the random value heuristic, which orders the variables randomly and does not imposes a partial order on the domain.

In what follows, we describe the *variable and value ordering* heuristics that we have specified in our search strategy. They are based on domain knowledge from a particular application. Of course, the closer the heuristic is to the objective function, the larger the subtree that can be pruned. Therefore, the more nodes pruned, the more efficient the search is. In our work, the best solution is the one that minimizes the query cost subject to space and maintenance cost constraints. In our model, the query cost reflects the local processing cost and the communication cost. As mentioned before, the communication cost is considered to be the dominant factor in a distributed environment. For this purpose, we apply the following heuristic: A view is preferably to be placed (materialized) closest to where it is frequently accessed. In the pseudo-code below, we describe our heuristic.

for *each* (v_i, s_j) *in* $V(G)$ **do**
  if $size(v_i) \leq Sp_{max_j}$ *and* $Mc(v_i, s_j) \leq U_{Max}$ **then**
    // *compute the benefit of materializing the view* v_i *on site* s_j
    $Benefit(v_i, s_j) = CCost(v_i, s_{origin} \rightarrow s_q) - CCost(v_i, s_j \rightarrow s_q)$
  end if
end for

Note that $CCost(v_i, s_{origin} \rightarrow s_q)$ is the cost of answering v_i at s_q by using the data origin and $CCost(v_i, s_j \rightarrow s_q)$ is the cost of answering v_i at s_q by using the data from the site s_j. This cost reflects the communication costs for transferring the data.

In the constraint solver, branching are applied to decision variables. In our constraint optimization model, the materialization variable $Mat(v_i, s_j)$ is the decision variable since the aim of the problem that we address in this work is to decide which views to materialize at what site of the network. Therefore, the variable selector has to start by instantiating the variables $Mat(v_i, s_j)$ corresponding to the views with highest benefit. For this purpose, we sort the views in $V(G)$ according to their *Benefit* in descending order (as it is presented below). We iterate over the sorted set starting with the views which have the highest benefit and we store them according to this order in the variable MVS.

//*sort according to the Benefit in descending order*
$VSSort = SortViewsSites(Benefit)$
for *each* (v_i, s_j) *in* $VSSort$ **do**
    $MVS = MVS \cup \{Mat(v_i, s_j)\}$
end for

Then, the variable selector will choose the materialization variable $Mat(v_i, s_j)$ in the order they appear in MVS. Once the variable has been selected, the value selector will assign the variable to its highest value: $max(d_{Mat(v_i, s_j)})$. As this way, the view v_i is considered as materialized at site s_j. By defining these heuristics in the search strategy, we expect that time and memory that the constraint solver incurs to find near-optimal solutions and the optimal solution will be significantly reduced since a large number of nodes in the search tree will be pruned.

6   Experimental Evaluation

This section present the results of a set of experiments designed to evaluate the effects of our approach on query processing time. The application takes as input a distributed scenario consisting of computer nodes with different resource constraints where each site (i.e., computer node) has its own query workload. As output, it produces the set of recommended views and sites at which they should be materialized. The distributed AND-OR view graph model, defined in Sect. 3, has been used as a tool to compactly represent alternative query plans, find commonalities among the queries and exploit materialized views whenever cost-effective.

Our approach uses a set of pruning heuristics, that reduce the solution space modeled by the graph, and search strategies which let the constraint solver converge quickly, to further bolster the claim that our approach is applicable to large-scale distributed environments.

The rest of this section is organized as follows. In Sect. 6.1, we describe our experimental setup, and the approach that we have used for comparison. In Sect. 6.2, we first study the effectiveness of the heuristics and search strategies that we have proposed in the previous section to trade off completeness for efficiency of the search. Then, we investigate the influence of resource limits on performance while varying the storage space and maintenance cost constraints. Next, we study the scalability of our approach for large query workloads as well as for complex scenarios by an increasing site numbers. Finally, we summarize the performance results in Sect. 6.3.

6.1   Experimental Setup

We have implemented our approach and compared it with the one presented in [7] that uses a genetic algorithm to compute the views and the sites to materialize them on. The study in [7] was chosen for comparison since it is the only effective and practical approach for large-scale distributed scenarios.

The setting that we have used in our experiments consists of a scenario simulation including a set of sites with different resource constraints (e.g., CPU, I/O, storage space capacity and bandwidth) and is additionally constrained by a maximum global maintenance cost. Each site query and update numerous relations on different sites. The query workload associated to each site is defined over the schema of the TPC-H benchmark [2]. The complexity of the query workload is important since query plans contain several join operations. The queries of the workload are randomly distributed over the network so that each *query initiator* has an associated query workload. The dataset is obtained by using the TPC-H relations which contain millions of tuples. The frequencies for access and update are randomly assigned based on a uniform distribution. We used the latest powerful version of CHOCO [1] to solve the distributed view selection problem (DVSP) as a constraint optimization problem (COP). Because the DVSP has been studied under resources constraints, we incorporate space and maintenance cost constraints into the genetic algorithm presented in [7]. All the methods are implemented in Java and all the experiments were carried out on a Quad-Core AMD Opteron(tm) Processor 8384 CPU @ 2,693 GHz machine running with 64 GB of RAM (the JVM was given 16 GB) and CentOS 5.9.

The performance of our approach was evaluated by measuring the gain in solution quality obtained by the materialized views. To evaluate the performance of distributed view selection methods, we measure the following metrics.

Solution Quality. The approach performance is evaluated by measuring the gain in solution quality obtained by view selection and data placement. The solution quality results from appraising the quality of the obtained set of views and sites at which they should be materialized in terms of cost savings. In the experimental results, the solution quality denoted by Q_s is computed as follows:

$$Q_s = \frac{WM - \sum_{(v_i, s_j) \in Q(G)} \left(f_q(v_i) * Qc(v_i, s_j) \right)}{WM - ALLM}$$

where WM and $AllM$ denotes respectively the "WithoutMat" approach and "AllMat" approach that are used as a benchmark for our normalized results. The "WithoutMat" approach does not materialize views and always recomputes queries whereas the "AllMat" approach materializes the result of each query of the workload at the site to whom it is associated this query. As defined in Sect. 4, $Qc(v_i, s_j)$ is the query cost corresponding to the view (v_i, s_j) and $f_q(v_i)$ is the query frequency. Note that the higher the Q_s-value , the better solution quality.

Space Constraint. The search space in this study is the total number of view combinations that meet the space constraint. The difficulty of solving the problem may depend on the relative magnitude of the storage space limits as compared with the size of the queries of workload. Following a common practice in literature [18], the storage space capacities are computed as follows.

$$Sp_{max_{s_i}} = \alpha * Sp_{AllM_{s_i}}$$

where $Sp_{max_{s_i}}$, defined in Sect. 4, is the maximum storage space corresponding to site s_i and $Sp_{AllM_{s_i}}$ is the size of the whole workload associated to site s_i. α is a constant where $\alpha \leq 1$ and $\alpha \succ 0$. We assume the case where the problem is studied under restrictive constraints as well as the case where we relax the space constraints for each site.

Maintenance Cost Constraint. The aim is to select a set of views and place them on the appropriate sites, while meeting the maintenance cost constraint. Similar to [18], we evaluate this constraint as follows.

$$U_{max} = \beta * Mc_{AllM}$$

where U_{max} is total view maintenance cost limit as defined in Sect. 4, Mc_{AllM} is the total maintenance cost when the result of each query of the workload is materialized and β is a constant. The value of β was set similar to α (see above).

Timeout Condition. The constraint solver provides ways to limit the tree search regarding different criteria. These limits have to be specified before the resolution. In our experiments, we use the two following timeout conditions.

– *Timeout(first-solution)*: The search stops when a solution is found.
– *Timeout(timelimit)*: Performing the search until reaching a search time limit *timelimit*. Observe that our approach is guaranteed to have some recommended views and sites to materialize them on at any time $T \geq T_{FS}$ (where T_{FS} is the time needed to find the first solution to the problem).

The search space in a constraint programming approach is explored to find feasible solutions and optimal solutions. However, the constraint solver may require an important amount of time to find the optimal solution to the DVSP, especially in large-scale distributed environments. In order to allow a fair comparison with the competitive approach [7], that uses a genetic algorithm to solve the DVSP, and since our approach is able to provide a solution at any time, the CHOCO solver was left to run until the convergence of the genetic algorithm in the following experiments. More precisely, the *timeout* condition was set to T_{GA} that is the time required by the genetic algorithm to solve the problem. Because our approach is known to be a powerful method for modeling and solving combinatorial optimization problems and our designed search strategies aim to speed up the search to optimal solutions, we expect to achieve a high solution quality when the genetic algorithm converges.

6.2 Impact of Heuristics and Optimizations

We study the impact of Join Ordering Heuristic (JOH), Site Selection Heuristic (SSH) and Variable and Value Ordering heuristics (VVO) on the search space explored by our approach which models the DVSP as a COP and by the genetic algorithm (GA). Two simple and small scenarios (Scenario 1 and Scenario 2) suffice to illustrate this.

The scenario 1 contains 5 sites which different constraints for CPU, I/O and network bandwidth. Two sites behave as *query initiator* ($|S_{QI}|$=2) and one site contains the original contents ($|S_{DO}|$=1). For each query initiator, it is associated a workload of 5 queries ($|Q_{S_{QI}}|$=5). ALL the sites may serve the role of *storage provider* or *query evaluator*. The roles that can be assigned to the sites of the network are explained in Sect. 3.1. Whereas the setting of the scenario 2 consists of 10 sites, each of which can evaluate any query of the workload or materialize pre-computed results. For this scenario, we assume that $|S_{QI}|$=4, $|S_{DO}|$=2 and $|Q_{S_{QI}}|$=5.

To study the benefit of our designed heuristics, we attempted to evaluate the reduction in the search space and compare the solution quality found by the genetic algorithm and our approach where the *timeout* is set to T_{GA} (COP) and T_{FS} (COP firstSol). We also compare the time that it takes to (i) solve the DVSP as a COP by the CHOCO solver and (ii) the convergence of the genetic algorithm (GA). In the following experiments, α and β, which define respectively the storage space and the view maintenance cost limits, was set to 0.4 and 0.6.

Impact of JOH. To evaluate the effectiveness of the JOH heuristic, we built different experiments with various values of K over]0,1]. The reason of varying the parameter K is explained in Sect. 5. In the case where K=1, all possible execution plans for each query of the workload are considered. The smaller the K-value, the more eliminate costly query execution strategies. Table 1 reports the reduction in the size of the graph for scenario 1 and scenario 2, as expressed by the total number of operation and equivalence nodes (views). The number of views and query plans is significantly reduced as the value of K decreases since we consider less query plan alternatives.

Figure 8a and b show respectively the solution quality of our approach and the one of the genetic algorithm over scenario 1 and scenario 2. We can observe that the reduction in the search space of candidate views conducts to a reduction in the solution quality. This confirms the intuition that considering only a subset of

Table 1. Comparing the sizes of the distributed AND-OR view graph over the *JOH*

K	Number of Eq-Nodes (views)	Number of Op-Nodes	TOTAL
Scenario 1			
0.25	105	832	937
0.5	120	1037	1157
0.75	122	1047	1169
1	135	1362	1497
Scenario 2			
0.25	350	7367	7717
0.5	360	7567	7927
0.75	363	7635	7998
1	400	10967	11367

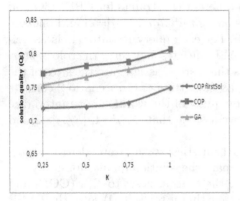

(a) Solution quality over Scenario 1

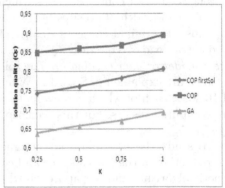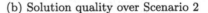

(b) Solution quality over Scenario 2

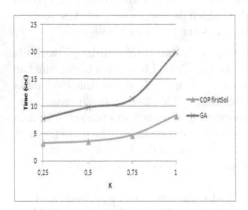

(c) Execution time over Scenario 1

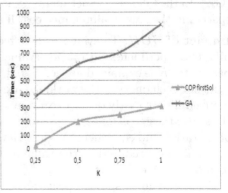

(d) Execution time over Scenario 2

Fig. 8. Solution quality and execution time while varying *K*

possible execution strategies lead to miss global optimal plans. However, the time needed to solve the problem by our approach and the genetic algorithm is significantly reduced for the instances with a relatively large reduction in the search space, as can be seen in Fig. 8c. The same trend was observed for scenario 2 (see Fig. 8d). Therefore, the small reduction in solution quality may be inconsequential compared to the significant reduction in execution time and search space (and hence memory needs).

Another important remark is that our approach achieves always significant gains in the solution quality compared to the genetic algorithm. This is even more pronounced in for the Scenario 2 (larger number of sites). Indeed the quality of the first solution found by the constraint solver is better that the one returned by the genetic algorithm.

Table 2. Comparing the sizes of the distributed AND-OR view graph over the SSH

S	Number of Eq-Nodes (views)	Number of Op-Nodes	TOTAL
Scenario 1			
S_{QI}	38	112	150
$S_{QI} \cup S_{DO}$	65	303	368
$S_{QI} \cup S_{NE}$	75	481	556
$S_{QI} \cup S_{DO} \cup S_{NE}$	97	911	1008
All sites	135	1362	1497
Scenario 2			
S_{QI}	72	376	448
$S_{QI} \cup S_{DO}$	145	1221	1366
$S_{QI} \cup S_{NE}$	159	2038	2197
$S_{QI} \cup S_{DO} \cup S_{NE}$	232	4043	4275
All sites	400	10967	11367

Impact of SSH. This heuristic can also help in reducing the solution space by considering for each query of the workload only a subset of sites during the data placement. We conduct experiments where the recommended sites are (i) $S = S_{QI}$; (ii) $S = S_{QI} \cup S_{DO}$; (iii) $S = S_{QI} \cup S_{NE}$ (iiii) $S = S_{QI} \cup S_{DO} \cup S_{NE}$; (iiiii) $S = \cup_{i=1..n} s_i$ where n is the number of sites in the network. Recall for Sect. 5 that S_{QI}, S_{DO} and S_{NE} correspond respectively to the query initiators, the data origins and the most promising neighbors of query initiators and data origins. Table 2 show the number of equivalence and operation nodes as we change the type of sites that we consider during the data placement. Obviously it appears that the magnitude of the reduction in the search space depends on the set of recommended sites for evaluating or materializing views. Next, we examine the impact of this reduction on the solution quality and the execution time.

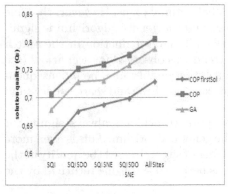

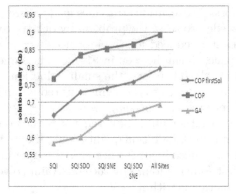

(a) Solution quality over Scenario 1   (b) Solution quality over Scenario 2

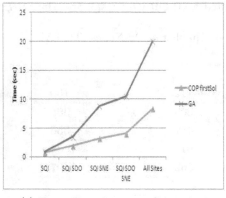

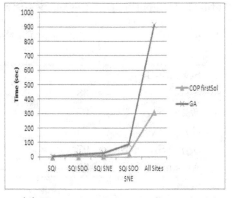

(c) Execution time over Scenario 1   (d) Execution time over Scenario 2

Fig. 9. Solution quality and execution time as a function of S

Figure 9a and b plots for Scenario 1 and Scenario 2 respectively the evolution of the quality of the solutions returned by our approach and the genetic algorithm.

We can observe that the quality decreases and tends to be relatively significant when we consider the first case in which we take into account only the query initiators since we discards a significant number of sites. For instance in Fig. 9b, the cost savings corresponding to the first case is degraded by approximately 13 % in comparison with the case where we consider all the sites of the network. However, this degradation is relatively small with respect to the fourth case ($S = S_{QI} \cup S_{DO} \cup S_{NE}$) which is less than 4 %.

From Fig. 9c and d, we note that the time to solve the DVSP decreases drastically while varying the recommended site set. Indeed, we observe for example in Fig. 9d that we could solve the problem within slightly over 90 sec where we consider the fourth case ($S = S_{QI} \cup S_{DO} \cup S_{NE}$) instead of 912 sec where we consider all the sites ($S = \cup_{i=1..n} s_i$). This represents a very important gain in the execution time of approximately 90 %. By considering the fourth case, we expect to reduce the search space and hence the evaluation time by several orders of magnitude with a relatively small reduction in the query performance. In all cases, our approach outperforms the genetic algorithm.

Impact of VV0. Here we study the impact of $VV0$ on the search space explored by the constraint solver and not on the solution space modeled by the distributed AND-OR view graph like with the previous heuristics. To evaluate this, we study the benefit that we bring to query evaluation and computing time when we use the VVO strategy and several combinations of heuristics denoted by $(JOH; VVO)$, $(SSH; VVO)$, $(JOH; SSH)$ and $(JOH; SSH; VVO)$. We also consider WH which represents the case where no heuristics is defined. Note that $VV0$ has no influence on the quality of the solution of the genetic algorithm as well as the time required by GA to converge. This can be explained by the fact that $VV0$ is defined only in the search strategy of the constraint solver. To ensure balance between solution quality and execution time, we set the parameters for JOH and SSH as follows: $k=0.5$ and $S = S_{QI} \cup S_{DO} \cup S_{NE}$.

The results on query performance are shown in Fig. 10a and b for Scenario 1 and Scenario 2 respectively. We can see that the quality of the different solutions of our approach is improved in the presence of $VV0$. This is because $VV0$ is a technique which helps the constraint solver to seek good (i.e. near-optimal) solutions in the beginning of the search. The direct descendants of each node in the search tree of the solver are totally ordered based on the variable and value ordering heuristic. Hence each move in the search tree is only performed if the resulting solution is better than the current solution. $VV0$ can also allows to build the search tree in the way that unfeasible solutions are eliminated subject to a subset of constraints (i.e. violation of space or maintenance constraints).

We can also observe from Fig. 10c and d that the time that a constraint solver incurs in the presence of a custom search for finding solutions is significantly reduced. This is because the variable and value ordering heuristics $VV0$ that we have defined in the search strategy reduce significantly the search space explored by the solver. Another important remark which is also valuable for the all experiments that we conducted so far, is that our approach provides good solution quality and can find a solution in a smaller amount of time. The performance differences are quite drastic with respect to the genetic algorithm. For scalability reasons, we have used the set of pruning heuristics JOH, SHH and VVO in the following experiments.

6.3 Performance Under Resource Constraints

To examine the impact of space and maintenance cost constraints on solution quality, we conduct experiments over a scenario of 100 sites with different constraints on CPU, IO and network bandwidth. The number of query initiators is equal to one-tenth of the total number of sites. Each query initiator issues ten queries so that the number of queries in the workload is equal to 100 queries. We assume that there is eight sites which behave as data origin. Each site of the network may serve the role of *storage provider* or *query evaluator*.

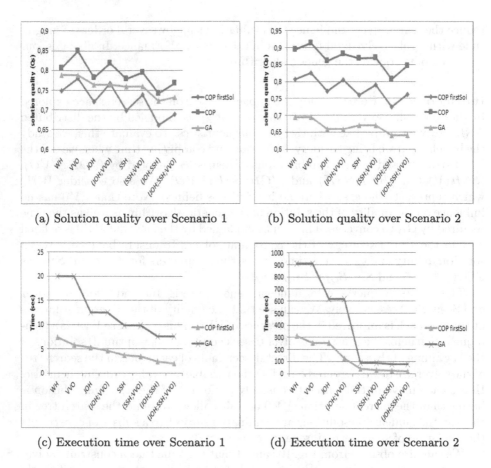

(a) Solution quality over Scenario 1   (b) Solution quality over Scenario 2

(c) Execution time over Scenario 1   (d) Execution time over Scenario 2

Fig. 10. Solution quality and execution time for different heuristic combinations

Performance While Varying the Space Constraint. In these experiments, we vary only the space constraint. Figure 11a investigates the influence of space limit on solution quality for each value of α which defines the storage space capacities (see Sect. 6.2). The values of α at each site are varied from 0.1 to 1. β that determines the view maintenance cost limit, is set to 0.6. We note that the quality of the solutions in terms of cost savings produced by the two methods improves when α increases, since there is storage space available for more views to be materialized. However, there is no improvement or very slight one in query performance from certain values of α because the maintenance cost constraint becomes the significant factor. Second, Fig. 11a shows that our approach provides the best gain in the solution quality for different values of α in comparison with the genetic algorithm.

Performance While Varying the Maintenance Cost Constraint. We discuss now the impact of the maintenance cost constraint on performance. Although the space constraint seems similar to the maintenance cost constraint, they have a

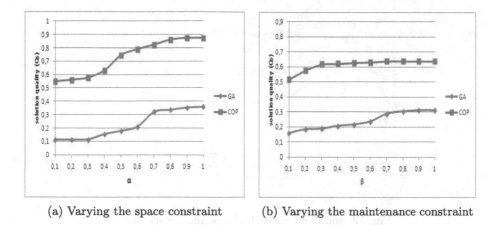

(a) Varying the space constraint   (b) Varying the maintenance constraint

Fig. 11. Performance under resource constraints

significant difference. The space occupied by a set of views always increases when a
new view is materialized, while the maintenance cost does not: it is possible for the
maintenance cost of a set of views to decrease after the addition of a new one. This
property is formally defined in [14]. In this experiment as it is shown in Fig. 11b,
the values of β are varied from 0.1 to 1 while α is set at each site to 0.4. Similarly to
the last experiment, we can see that we have better solutions when β increases since
the maintenance time window is available for making the materialized views up-to-
date. The performance stabilizes from certain values of β since the space constraint
becomes the dominant parameter. Our approach again performs consistently better than the other method.

6.4   Scalability of Our Approach

To evaluate the scalability of our proposed approach, we attempted to solve the
problem for instances with a larger number of queries in the workload. Next, we
present the results for large-scale distributed scenarios to show how our approach
scales well to a large number of sites.

Solution Quality on Large Query Workloads. Let us now evaluate the performance of our approach and the one of genetic algorithm on larger query workloads.
To this purpose, we generated a scenario of 100 sites where the query workload
ranges from 100 to 1000 queries. The number of query initiators is equal to one-
tenth of the total number of sites. While the number of data origin is equal to eight.
As in previous experiments, any site of the network may serve the role of *storage
provider* or *query evaluator*. For the space and maintenance cost constrains, $\alpha=0.4$
at each site of the network and $\beta=0.6$.

The experiment results are depicted in Fig. 12. We can make the following observations. Our approach provides better performances in terms of the solution quality

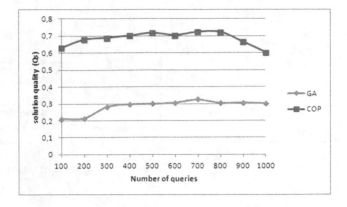

Fig. 12. Solution Quality for large query workloads

while varying the number of queries compared to the other algorithm. Indeed, our approach provides the better solution by a significant margin. The cost saving is up to 48 % more than the genetic algorithm. The quality of the solution is slightly decreasing with a workload of 1000 queries. The *timeout* condition is set close to 5 h. This is the time that the genetic takes to converge with a scenario of 100 sites and a workload of 1000 queries which correspond to thousands of views. We conclude that our approach scales well up to 1000 queries since the quality of the results is not significantly influenced by an increasing query number.

Solution Quality on Large-Scale Distributed Scenarios. We study the scalability of our approach for complex scenario which contains large number of sites. We run experiments while varying the number of sites from 100 to 1000 sites. Each query initiator poses ten queries. As in previous experiments, query initiators represent one-tenth of the total number of sites and the base relations are stored at

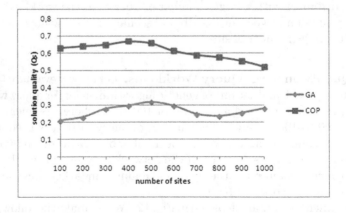

Fig. 13. Solution Quality for large-scale distributed scenarios.

eight different sites. The sites of the network are heterogeneous with respect to the resource constraints (e.g., CPU, I/O and network bandwidth) and may serve any role of those defined in Sect. 3. The DVSP is studied under space and maintenance constraints. For each site, α is set to 0.4 and for the maintenance cost constraint, β is set to 0.6.

The experiment results are shown in Fig. 13. We can observe that an increase in the number of sites does not seem to affect much the solution quality in terms of cost savings, which indicates the effectiveness of our approach in large-scale scenarios. Note that the *timeout* condition is slightly over 16 h which is the time that the genetic algorithm takes to converge for a scenario of 1000 sites. We once again see that our approach performs the best, finding significantly better quality of the obtained set of materialized views in terms of cost saving compared with the other method.

6.5   Experiment Conclusion

Our experiments show that our approach achieves significant performance gains. Indeed, our recommended views do improve the query evaluation with impressive cost saving factors subject to space and maintenance constraints. The JOH, SSH heuristics that we have designed are efficient and effective since they reduce significantly the search space and hence the density of the distributed AND-OR view graph with a relatively small reduction in the solution quality. The VVO heuristic that we have defined within the constraint solver helps the latter to reach high solution quality from the beginning of the search by discarding a large number of inferior solutions. The experiment results confirm our expectation that the set of pruning heuristics that we have proposed allows our approach to scale well with large query workloads up to 1000 queries as well as with complex scenarios which contains up to 1000 heterogeneous sites. Finally, we can conclude from the experimental evaluation that our constraint optimization approach outperforms the genetic algorithm by several orders of magnitude.

7   Concluding Remarks and Future Work

In this paper, we formalize and study the distributed view selection problem which is the combined problem of view selection and data placement. Our work explores the full potential of view materialization techniques for improving query performance. In this context, we address the problem of efficiently recommending a set of views and a set of sites to materialize them on, subject to view storage and view maintenance cost constraints. We have proposed a new approach which is based on the integration of constraint programming that is a well known solution method to the combinatorial optimization problems. To this purpose, the distributed view selection problem has been modeled as a constraint optimization problem. This study focuses on the construction of effective and efficient heuristics that reduce drastically the solution space to solve the problem with large-scale distributed scenarios, for which an early version of this work failed to produce any solution as they outgrow the available memory. We evaluate the performance of our approach over various experiments. Our experimental study corroborates our claim that our approach can significantly improve the query performance and by a drastic margin compared to the other method. The evaluation also confirms that our approach is applicable for large number of queries or sites. As a future work, we are planning to extend our constraint optimization model and adapt our approach to address the dynamic distributed view selection problem which is still an open issue. Computing which view has to be materialized at what machine in the cloud computing, in which large amounts of data, content and knowledge are being spread over the service providers' infrastructures, is another avenue of further research. It seems interesting to explore new costs models in this context. Indeed, [33] has been worked on a novel cost models that complement the existing materialized view cost models with a monetary cost component that is primordial in the cloud.

References

1. Choco, open-source software for constraint satisfaction problems. http://www.emn.fr/z-info/choco-solver
2. The TPC benchmark H (TPC-H). http://www.tpc.org/tpch/spec/tpch2.14.3.pdf
3. Bauer, A., Lehner, W.: On solving the view selection problem in distributed data warehouse architectures. In: SSDBM, pp. 43-51, Cambridge (2003)
4. Apt, K.: Principles of Constraint Programming. Cambridge University Press, New York (2003)
5. Baril, X., Bellahsene, Z.: Selection of materialized views: a cost-based approach. In: CAiSE, pp. 665–680, Klagenfurt (2003)
6. Bellahsene, Z., Cart, M., Kadi, N.: A cooperative approach to view selection and placement in P2P systems. In: Meersman, R., Dillon, T.S., Herrero, P. (eds.) OTM 2010. LNCS, vol. 6426, pp. 515–522. Springer, Heidelberg (2010)
7. Chaves, L.W.F., Buchmann, E., Hueske, F., Böhm, K.: Towards materialized view selection for distributed databases. In: Proceedings of the 12th International Conference on Extending Database Technology: Advances in Database Technology, EDBT 2009, pp. 1088–1099. ACM, New York (2009)
8. Derakhshan, R., Dehne, F.K., Korn, O., Stantic, B.: Simulated annealing for materialized view selection in data warehousing environment. In: Databases and Applications, pp. 89–94, (2006)
9. Derakhshan, R., Stantic, B., Korn, O., Dehne, F.K.H.A.: Parallel simulated annealing for materialized view selection in datawarehousing environments. In: ICA3PP, pp. 121–132, island of Cyprus (2008)
10. Deshpande, P.M., Ramasamy, K., Shukla, A., Naughton, J.F.: Caching multidimensional queries using chunks. In: SIGMOD Conference, pp. 259–270, Seattle (1998)
11. Du, W., Krishnamurthy, R., Shan, M.C.: Query optimization in heterogeneous dbms. In: Proceeding of VLDB, pp. 277–91, Vancouver (1992)
12. Gribble, S.D., Halevy, A.Y., Ives, Z.G., Rodrig, M., Suciu, D.: What can database do for peer-to-peer? In: WebDB, pp. 31–36, Santa Barbara (2001)
13. Gupta, H.: Selection of views to materialize in a data warehouse. In: ICDT, pp. 98–112, Delphi (1997)
14. Gupta, H., Mumick, I.S.: Selection of views to materialize under a maintenance cost constraint. In: Beeri, C., Bruneman, P. (eds.) ICDT 1999. LNCS, vol. 1540, pp. 453–470. Springer, Heidelberg (1998)

15. Gupta, H., Mumick, I.S.: Selection of views to materialize in a data warehouse. IEEE Trans. Knowl. Data Eng. **17**(1), 24–43 (2005)
16. Harinarayan, V., Rajaraman, A., Ullman, J.D.: Implementing data cubes efficiently. In: SIGMOD Conference, pp. 205–216, Montreal (1996)
17. Horng, J.-T., Chang, Y.-J., Liu, B.-J.: Applying evolutionary algorithms to materialized view selection in a data warehouse. Soft Comput. **7**(8), 574–581 (2003)
18. Kalnis, P., Mamoulis, N., Papadias, D.: View selection using randomized search. Data Knowl. Eng. **42**(1), 89–111 (2002)
19. Kalnis, P., Ng, W.S., Ooi, B.C., Papadias, D., Tan, K.L.: An adaptive peer-to-peer network for distributed caching of olap results. In: SIGMOD Conference, pp. 25–36, Madison (2002)
20. Kossmann, D.: The state of the art in distributed query processing. ACM Comput. Surv. **32**(4), 422–469 (2000)
21. Kossmann, D., Franklin, M.J., Drasch, G.: Cache investment: integrating query optimization and distributed data placement. ACM TODS **25**2000 (2000)
22. Kotidis, Y., Roussopoulos, N.: Dynamat: a dynamic view management system for data warehouses. In: SIGMOD Conference, pp. 371–382, Philadephia (1999)
23. Kumar, T.V., Kumar, S.: Materialized view selection using genetic algorithm. In: IC3, pp. 225–237 (2012)
24. Labio, W.J., Quass, D., Adelberg, B.: Physical database design for data warehouses. In Proceedings of the Thirteenth International Conference on Data Engineering, ICDE 1997, pp. 277–288. IEEE Computer Society, Washington, DC (1997)
25. Lee, M., Hammer, J.: Speeding up materialized view selection in data warehouses using a randomized algorithm. Int. J. Cooperative Inf. Syst. **10**(3), 327–353 (2001)
26. Ligoudistianos, S., Theodoratos, D., Sellis, T.: Experimental evaluation of data warehouse configuration algorithms. In: DEXA Workshop, pp. 218–223, Vienna (1998)
27. Mackert, L.F., Lohman, G.M.: R* optimizer validation and performance evaluation for local queries. In: Proceedings of the ACM SIGMOD International Conference on Management of Data, SIGMOD 1986, pp. 84–95. ACM, New York (1986)
28. Mami, I., Bellahsene, Z.: A survey of view selection methods. SIGMOD Record **41**(1), 20–29 (2012)
29. Mami, I., Bellahsene, Z., Coletta, R.: View selection under multiple resource constraints in a distributed context. In: Liddle, S.W., Schewe, K.-D., Tjoa, A.M., Zhou, X. (eds.) DEXA 2012, Part II. LNCS, vol. 7447, pp. 281–296. Springer, Heidelberg (2012)
30. Mami, I., Bellahsene, Z., Coletta, R.: A declarative approach to view selection modeling. Trans. Large-Scale Data Knowl. Centered Syst. **10**, 115–145 (2013)
31. Mami, I., Coletta, R., Bellahsene, Z.: Modeling view selection as a constraint satisfaction problem. In: Hameurlain, A., Liddle, S.W., Schewe, K.-D., Zhou, X. (eds.) DEXA 2011, Part II. LNCS, vol. 6861, pp. 396–410. Springer, Heidelberg (2011)
32. Mistry, H., Roy, P., Sudarshan, S., Ramamritham, K.: Materialized view selection and maintenance using multi-query optimization. In: SIGMOD Conference, pp. 307–318, Santa Barbara (2001)
33. Nguyen, T.V.A., Bimonte, S., d'Orazio, L., Darmont, J.: Cost models for view materialization in the cloud. In: EDBT/ICDT Workshops, pp. 47–54, Berlin (2012)
34. Özsu, M.T., Valduriez, P.: Principles of Distributed Database Systems, 3rd edn. Springer, New York (2011)
35. De Raedt, L., Guns, T., Nijssen, S.: Constraint programming for itemset mining. In: KDD, pp. 204–212, Las Vegas (2008)
36. Rossi, F., van Beek, P., Walsh, T.: Handbook of Constraint Programming (Foundations of Artificial Intelligence). Elsevier Science Inc., New York (2006)

37. Roy, P., Seshadri, S., Sudarshan, S., Bhobe, S.: Efficient and extensible algorithms for multi query optimization. In: SIGMOD Conference, pp. 249–260, Dallas (2000)
38. Scheuermann, P., Shim, J., Vingralek, R.: A data warehouse intelligent cache manager. In: VLDB, pp. 51–62, Bombay (1996)
39. Selinger, P.G., Astrahan, M.M., Chamberlin, D.D., Lorie, R.A., Price, T.G.: Access path selection in a relational database management system, pp. 23–34 (1979)
40. Steinbrunn, M., Moerkotte, G., Kemper, A.: Heuristic and randomized optimization for the join ordering problem. VLDB J. 6(3), 191–208 (1997)
41. Theodoratos, D., Ligoudistianos, S., Sellis, T.K.: View selection for designing the global data warehouse. Data Knowl. Eng. 39(3), 219–240 (2001)
42. Theodoratos, D., Sellis, T.K.: Data warehouse configuration. In: VLDB, pp. 126–135, Athens (1997)
43. Ye, W., Gu, N., Yang, G., Liu, Z.: Extended derivation cube based view materialization selection in distributed data warehouse. In: Fan, W., Wu, Z., Yang, J. (eds.) WAIM 2005. LNCS, vol. 3739, pp. 245–256. Springer, Heidelberg (2005)
44. Yu, J.X., Yao, X., Choi, C.H., Gou, G.: Materialized view selection as constrained evolutionary optimization. IEEE Trans. Syst. Man Cybern. Part C 33(4), 458–467 (2003)
45. Zhang, C., Yang, J.: Genetic Algorithm for materialized view selection in data warehouse environments. In: Mohania, M., Tjoa, A.M. (eds.) DaWaK 1999. LNCS, vol. 1676, pp. 116–125. Springer, Heidelberg (1999)
46. Zhang, C., Yao, X., Yang, J.: An evolutionary approach to materialized views selection in a data warehouse environment. IEEE Trans. Syst. Man Cybern. Part C 31(3), 282–294 (2001)
47. Zhou, J., Larson, P-Å, Goldstein, J., Ding, L.: Dynamic materialized views. In: ICDE, pp. 526–535, Istanbul (2007)

On the Selection of SPARQL Endpoints to Efficiently Execute Federated SPARQL Queries

Maria-Esther Vidal[1]([✉]), Simón Castillo[1], Maribel Acosta[2],
Gabriela Montoya[3], and Guillermo Palma[1]

[1] Universidad Simón Bolívar, Caracas, Venezuela
{mvidal,scastillo,gpalma}@ldc.usb.ve
[2] Institute AIFB, Karlsruhe Institute of Technology, Karlsruhe, Germany
maribel.acosta@kit.edu
[3] University of Nantes, Nantes, France
Gabriela.Montoya@univ-nantes.fr

Abstract. We consider the problem of source selection and query decomposition in federations of SPARQL endpoints, where query decompositions of a SPARQL query should reduce execution time and maximize answer completeness. This problem is in general intractable, and performance and answer completeness of SPARQL queries can be considerably affected when the number of SPARQL endpoints in a federation increases. We devise a formalization of this problem as the Vertex Coloring Problem and propose an approximate algorithm named *Fed-DSATUR*. We rely on existing results from graph theory to characterize the family of SPARQL queries for which *Fed-DSATUR* can produce optimal decompositions in polynomial time on the size of the query, i.e., on the number of SPARQL triple patterns in the query. *Fed-DSATUR* scales up much better to SPARQL queries with a large number of triple patterns, and may exhibit significant improvements in performance while answer completeness remains close to 100 %. More importantly, we put our results in perspective, and provide evidence of SPARQL queries that are hard to decompose and constitute new challenges for data management.

1 Introduction

Over the past decade, the number of Linked Data sets in the Linking Open Data cloud has exploded as well as the number of SPARQL endpoints that access these datasets [25]. As more Linked Open Data becomes available, applications from different domains are frequently developed, and queries that require gathering data from several Linked Data sets are more likely everyday. Linked Data applications include [9]: TELEIOS[1] relies on Linked Geospatial Data to uncover hidden patterns in earth observations; OpenPHACTS[2] links pharmacological data in order to provide support for the discovery of new patterns and connections;

---

[1] http://www.earthobservatory.eu/.
[2] https://www.openphacts.org/.

© Springer-Verlag Berlin Heidelberg 2016
A. Hameurlain et al. (Eds.): TLDKS XXV, LNCS 9620, pp. 109–149, 2016.
DOI: 10.1007/978-3-662-49534-6_4

and GeoKnow[3] integrates geo-spatial knowledge with the Linked Data Web for data exploration and curation. As a result, given the vast amount of Linked Data sets and the diversity of applications, federated SPARQL query engines need to provide a unify querying interface to large federations of SPARQL endpoints, and execute SPARQL queries against the federation in a way that execution time is minimized while the query answer completeness is maximized.

So far, several approaches have addressed the problem of decomposing a SPARQL query into subqueries that can be executed on the Web of Data [1, 3, 13, 27]. Some approaches for the source selection decision rely on statistics collected from the sources [13] or simply consider all possible subqueries and choose the most promising ones [3]. Others implement heuristic-based strategies to identify the subqueries that can be executed by the available SPARQL endpoints [1, 27]. For example, FedX [27] is a rule-based system able to generate left-linear plans comprised of subqueries that can be exclusively answered by one SPARQL endpoint (*exclusive groups*); FedX does not derive the query decomposition decision on knowledge about schema alignments or data distributions. ANAPSID [1] resorts to source descriptions to determine the SPARQL endpoints that can answer a SPARQL triple pattern. All the triple patterns that can be executed the same set of SPARQL endpoints are grouped together in a way that the number of operations done by the selected endpoints are maximized while the size of intermediate results is minimized. SPLENDID [10] exploits statistics that describe RDF data accessible from the federation of SPARQL endpoints to perform the source selection and query optimization tasks. DAW [24] exploits information encoded in index-based descriptions of the data accessible via SPARQL endpoints, to identify execution plans that reduce the gathering of replicated data from the selected SPARQL endpoints. Finally, HiBISCuS [23] also resorts to index-based structures to discard the SPARQL endpoints that do not contribute to the final answer. Although these federated approaches may effectively address the source selection problem, performance may be deteriorated whenever SPARQL queries with a large number of triple patterns are executed or large intermediate results are retrieved from the SPARQL endpoints.

In this paper, we study the characteristics of SPARQL queries for which existing federated SPARQL query engines may perform poorly, and the complexity of decomposing these queries into efficient and effective plans. We conduct our study from a theoretical and empirical point of view, and provide formal proofs that reveal a theoretical justification of the behavior shown by state-of-the-art engines in a class of SPARQL queries. The main idea of our approach is to cast the SPARQL federated query decomposition problem into the Vertex Coloring Problem such that colorings correspond to decompositions of a SPARQL query against the SPARQL endpoints of a federation. Additionally, building on existing results from graph theory, we identify properties of the SPARQL queries that ensure *optimality* of the query decomposition, i.e., the partition of the original query into subqueries executable on available SPARQL endpoints whose *execution cost* is minimal and the percentage of *answer completeness* is

---

[3] http://geoknow.eu/.

100 %. *Fed-DSATUR* extends DSATUR [4], a coloring approximate algorithm that implements a greedy iterative strategy to color each node once, while it follows a heuristic to color first the nodes with a reduced number of possible colors. To experimentally evaluate *Fed-DSATUR*, we study the topology of the graphs generated from both FedBench queries [26] and additional queries comprised of a large number of triple patterns, and compare the performance of state-of-the-art engines during the execution of these queries. Our experimental results reveal that for SPARQL queries whose optimal solutions can be identified in polynomial time, existing engines perform similarly. In contrast, for SPARQL queries that do not meet this property, dissimilar behaviors are observed, raising new challenges to state-of-the-art federated SPARQL query engines.

In summary, we make the following crisp contributions to the problem of federated SPARQL query decomposition:

- Formalization of the Federated SPARQL Query Decomposition Problem (FSQD) and characterization of tractability conditions.
- Mapping of the Federated SPARQL Query Decomposition Problem (FSQD) into the Vertex Coloring Problem. Definition of the optimization criteria in a way that the generated query decompositions can reduce execution time while answer completeness is maximized.
- An approximate solution named *Fed-DSATUR* that extends the greedy algorithm DSATUR [4] to solve the problem of source selection and query decomposition on federations of SPARQL endpoints. *Fed-DSATUR* does not rely on statistics, indices, or any kind of estimates during source selection and query decomposition. Therefore, *Fed-DSATUR* adapts the selection criteria to the current conditions of the RDF datasets, e.g., the existence or not of RDF triples of a given predicate in an RDF dataset.
- An empirical evaluation of *Fed-DSATUR* and existing approaches on diverse instances of the problem. Queries and datasets from the FedBench benchmark are used in the study. We configure two federations of SPARQL endpoints to access the FedBench data collections. In both federations RDF triples are fragmented in datasets accessible via SPARQL endpoints. The goal of the study is to evaluate the impact of fragmenting data across several RDF datasets on the behavior of the studied federated SPARQL query engines. Furthermore, additional *complex* SPARQL queries which comprise large number of triple patterns and SPARQL operators are evaluated. We aim at evaluating the behavior of federated SPARQL query engines in queries with large spaces of potential query decompositions; some of these SPARQL queries raise new challenges to the evaluated approaches.

The rest of the paper is as follows: Sect. 2 gives a motivating example. Section 3 summarizes the terminology and concepts required to understand our proposed approach. Section 4 presents the Federated SPARQL Query Decomposition Problem (FSQD) and the *Fed-DSATUR* approximate algorithm. Experimental results are reported in Sect. 5. Section 6 summarizes the related work. Finally, we conclude in Sect. 7 with an outlook to future work.

2 Preliminaries

2.1 RDF and SPARQL

An RDF graph [21] is a set of RDF triples (s, p, o), where s represents a subject, p is a predicate, and o is an object. Formally, an RDF graph T is as follows:

$$T \subseteq (U \cup B) \times U \times (U \cup B \cup L)$$

- U: An infinite set of URI references;
- B: An infinite set of blank nodes; and
- L: An infinite set of RDF literals.

A SPARQL query Q is represented by a set of the basic graph patterns $BGPs$ in the WHERE clause of Q; each BGP is a set of triple patterns, and $BGPs$ can be connected by OPTIONALs or UNIONs in Q. An exact star $ES(P, ?X)$ of triple patterns of a basic graph pattern P on a variable $?X$ is as follows [29]:

- $ES(S, ?X)$ is a triple pattern in S of the form $\{s\ p\ ?X\ \}$ or $\{?X\ p\ o\}$ such that, $s \neq ?X$, $p \neq ?X$ and $o \neq ?X$.
- $ES(S, ?X)$ is the union of two exact starts, $ES1(S, ?X)$ and $ES2(S, ?X)$, such that they only share the variable $?X$, i.e., $var(ES1(S, ?X)) \cap var(ES2(S, ?X)) = \{?X\}$.

Consider the following LS5 SPARQL query: "Drugs and their components' url and image", see Listing 1.1.

Listing 1.1. Query LS5 from FedBench [26]

```
1   PREFIX rdf: <http://www.w3.org/1999/02/22-rdf-syntax-ns#>
2   PREFIX drugbank: <http://www4.wiwiss.fu-berlin.de/drugbank/resource/drugbank/>
3   PREFIX purl:<http://purl.org/dc/elements/1.1/>
4   PREFIX bio2rdf:<http://bio2rdf.org/ns/bio2rdf#>
5   SELECT ?drug ?keggUrl ?chebiImage WHERE {
6       ?drug rdf:type drugbank:drugs .
7       ?drug drugbank:keggCompoundId ?keggDrug .
8       ?drug drugbank:genericName ?drugBankName .
9       ?keggDrug bio2rdf:url ?keggUrl .
10      ?chebiDrug purl:title ?drugBankName .
11      ?chebiDrug bio2rdf:image ?chebiImage
12  }
```

LS5 comprises one basic graph pattern BGP of six triple patterns. This BGP can be decomposed into one subquery composed of the triple pattern in line 9, and into two star groups: (i) One star subquery is on the variable $?drug$ and is composed of three triple patterns in the lines 6, 7, and 8; and (ii) The other star subquery comprises the triple patterns in lines 10 and 11 and is on the variable $?chebiDrug$. Different studies reported in the literature [17,29, 31] suggest that query plans composed of small-sized star-shaped groups can be effectively executed on existing RDF triple stores. We rely on these results and propose a query decomposition technique able to identify stars groups as subqueries that can be executed against the SPARQL endpoints of a federation. The proposed decomposition techniques benefit the generation of subqueries that correspond to maximal star-shaped subqueries whenever the non-selected SPARQL endpoints do not affect the completeness of the query answer.

2.2   Federations of SPARQL Endpoints

A federation of SPARQL endpoints is a set of RDF datasets that can be accessed via SPARQL endpoints. A SPARQL endpoint is a Web service that provides a Web interface to query RDF data following the SPARQL protocol[4]. RDF datasets comprise sets of RDF triples; predicates of these triples can be from more than one Linked Open Vocabulary[5], e.g., FOAF[6] or DBpedia ontology. Additionally, proprietary vocabularies can be used to describe the RDF resources of these triples, and controlled vocabularies as VoID[7], can be used to describe the properties of the RDF data accessible through a given SPARQL endpoint. In this work, we assume no information about the properties of the RDF data accessible through a SPARQL endpoints is available; only, the URL is provided as a description of the endpoint.

Table 1 illustrates a federation comprised of 26 SPARQL endpoints which provide access to 26 RDF datasets. These RDF collections are part of Fed-Bench [26] which is the only benchmark available to evaluate performance and behavior of the query processing techniques implemented in the existing federated query engines. We use FedBench in our running examples and in the empirical evaluation of our approach. FedBench comprises three data collections:

Cross Domain Collection: Data from different linked domains, e.g., movies (LMDB)[8], DBpedia[9], GeoNames[10], news from the New York Times (NYT)[11], information from the Semantic Web conferences (SW Dog Food)[12], and music from Jamendo[13].

Life Science Collection: Biomedical data with drugs and targets from Drugbank[14] and DBpedia, genes and genomes from KEGG[15], and molecular entities on chemical components from ChEBi[16].

SP²Bench Data Collection: Synthetic data divided into different smaller datasets according to the type of the data.

Additionally, FedBench provides four sets of SPARQL queries designed to stress query answer completeness and query execution performance. In total the benchmark comprises 39 SPARQL queries:

---

[4] http://www.w3.org/TR/rdf-sparql-protocol/.
[5] http://lov.okfn.org/dataset/lov.
[6] http://xmlns.com/foaf/spec/.
[7] http://www.w3.org/TR/void/.
[8] http://www.linkedmdb.org/.
[9] http://dbpedia.org/About.
[10] http://www.geonames.org/.
[11] http://data.nytimes.com/.
[12] http://data.semanticweb.org/.
[13] http://dbtune.org/jamendo/.
[14] http://www.drugbank.ca/.
[15] http://www.genome.jp/kegg/.
[16] http://www.ebi.ac.uk/chebi/.

Table 1. Running example: federation of SPARQL endpoints. Fed_1 is a federation of 26 SPARQL endpoints to access FedBench RDF datasets

RDF dataset name	# Endpoints	# Triples per endpoint
NY Times (NYT)	1	314 k
LinkedMDB (LMDB)	1	6.14 M
Jamendo	1	1.04 M
Geonames (eleven data sets)	10	9.9 M
	1	7.98 M
SW Dog Food (SWDF)	1	84 k
KEGG	1	10.9 M
Drugbank	1	517 k
ChEBi	1	4.77 M
SP^{2}B-10M	1	10 M
DBpedia		
Infobox types	1	5.49 M
Infobox properties	1	10.80 M
Titles	1	7.33 M
Articles categories	1	10.91 M
Images	1	3.88 M
SKOS categories	1	2.24 M
Other	1	2.45 M

Cross Domain (CD): Seven full SPARQL queries[17] against RDF datasets in the Cross Domain Collection.

Life Science (LS): Seven full SPARQL queries against RDF datasets in the Life Science Collection.

Linked Data (LD): Eleven basic graph patterns SPARQL queries against RDF datasets in the Cross Domain and Life Science Collections.

SP^2: 14 full SPARQL queries against RDF datasets in the SP^2 Bench Data Collection.

Queries against federations of SPARQL endpoints are posed through federated SPARQL query engines. Figure 1 presents a generic architecture of a federated SPARQL query engine. This architecture is based on the mediator and wrapper architecture [30]. Light-weight wrappers translate SPARQL subqueries into calls to the SPARQL endpoints as well as convert endpoint answers into the query engine internal structures. The mediator is composed of three main components: *(i) Source Selection and Query Decomposer:* Decomposes user queries into multiple simple subqueries, and selects the endpoints that are capable of executing each subquery. Simple subqueries comprise a list of triple patterns that

---

[17] SPARQL queries with different SPARQL operators, e.g., UNION or OPTIONAL.

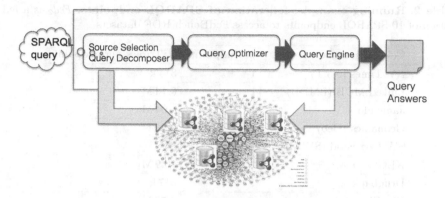

Fig. 1. Typical federated SPARQL query engine architecture. Main components: *Source selection and query decomposition* identifies SPARQL endpoints to execute the subqueries of original query; *Query optimizer* generates physical plans; and *Query engine* executes the physical plan against the selected SPARQL endpoints

can be evaluated against at least one endpoint. *(ii) Query Optimizer:* Identifies execution plans that combine subqueries and benefits the generation of specific physical plans that exploit the properties of the physical operators implemented by the query engine. Statistics about the distribution of values in the different datasets may be used to identify the best combination of subqueries as well as the shape of the plan that will ensure an efficient execution of the query. *(iii) Query Engine:* Implements different physical operators to combine tuples from different endpoints. We focus on the *Source Selection and Query Decomposer* component and propose techniques able to reduce execution time without affecting query answer completeness. Because properties of the SPARQL endpoints can be collected on the fly by executing SPARQL ASK queries, federated SPARQL query engines can implement adaptivity at the level of source selection and query decomposition, i.e., these are *adaptive planning engines*.

3   Motivating Example

We motivate our work by observing how the performance and answer completeness of queries executed against federations of SPARQL endpoints are impacted by the type of decomposition. An experiment was set up in order to evaluate the performance and answer completeness of different query decompositions of the query CD6 (Cross Domain) of FedBench. We fragmented the FedBench collections[18] as reported in Table 2. The federated query engine ANAPSID [1] (Version May 2014) was used to evaluate CD6 as well as the different decompositions. Figure 3 presents CD6 and the endpoints that access RDF datasets comprised of RDF triples with the predicates used in the query. As can be observed in Fig. 3,

---

[18] http://iwb.fluidops.com:7879/resource/Datasets, November 2011.

Table 2. Running example: federation of SPARQL endpoints. Fed_2 is a federation of 10 SPARQL endpoints to access FedBench RDF datasets

Collection name	# Endpoints	# Triples per endpoint
NY Times (NYT)	1	314 k
Linked MDB (LMDB)	1	6.14 M
Jamendo	1	1.04 M
Geonames (Geo)	1	9.9 M
SW Dog Food (SWDF)	1	84 k
KEGG	1	10.9 M
Drugbank	1	517 k
ChEBi	1	4.77 M
DBpedia	1	43.1 M
SP^{2}B-10M (Bibliographic)	1	10 M

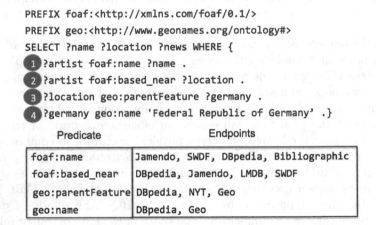

Fig. 2. Running example cross domain FedBench query CD6 and relevant endpoints per RDF predicate of CD6 triple patterns

seven different SPARQL endpoints can be used to execute the triple patterns of CD6; these SPARQL endpoints correspond to our baseline.

Further, Fig. 3(a)–(c) show three decompositions for contacting the selected endpoints and executing CD6. For each decomposition we present the number of subqueries, and the triple patterns assigned to the subqueries as well as the endpoints selected for each subquery. Further, we summarize the total number of endpoints assigned to the decomposition as the Number of Contacts to Endpoints (NCE); the Number of Non-Selected Endpoints (NNME) corresponds to the number of endpoints that even being relevant for a triple pattern are not selected for the subquery where the triple pattern is assigned in the decomposition. For example, decomposition D_1 comprises four subqueries, one per triple pattern;

SubQuery (SQ)	Triple Patterns	Endpoints
SQ1	1	Jamendo, SWDF, DBpedia, Bibliographic
SQ2	2	LMDB, SWDF, DBpedia, Jamendo
SQ3	3	DBpedia, NYT, Geo
SQ4	4	DBpedia, Geo

Number of Contacts to Endpoints (NCE): 13
Number of Non-Selected Endpoints (NNSE): 0

(a) Decomposition D_1

SubQuery (SQ)	Triple Patterns	Endpoints
SQ1	1 2	SWDF, Dbpedia, Jamendo
SQ3	3 4	DBpedia, Geo

Number of Contacts to Endpoints (NCE): 5
Number of Non-Selected Endpoints (NNSE): 3

(b) Decomposition D_2

SubQuery (SQ)	Triple Patterns	Endpoints
SQ1	1	Jamendo, SWDF, DBpedia, Bibliographic
SQ2	2	LMDB, SWDF, DBpedia, Jamendo
SQ3	3 4	DBpedia, Geo

Number of Contacts to Endpoints (NCE): 10
Number of Non-Selected Endpoints (NNSE): 1

(c) Decomposition D_3

Fig. 3. Running example: decompositions for FedBench CD6 query. Each query decomposition is defined in terms of subqueries, Number of Contacts to Endpoints (NCE), and Number of Non-Selected Endpoints (NNME). NCE corresponds to the sum of the number of relevant endpoints assigned to each subquery. NNME represents the number of endpoints that are relevant for a triple pattern but that were not selected in the decomposition

all the endpoints that define the corresponding predicate of each triple pattern are selected; thus, Number of Contacts to Endpoints (NCE) is the sum of the endpoints contacts per subquery, i.e., 13; the Number of Non-Selected Endpoints (NNME) is 0. On the other hand, decomposition D_2 is composed of two subgoals

Table 3. Running example: Execution Time seconds (ET) and Percentage of Answers (PA) for CD6. Given decompositions D_1, D_2, and D_3 of CD6, queries are executed on ANAPSID (Version May 2014) against 10 SPARQL Virtuoso endpoints that access the FedBench collections (*Fed₂*)

Decompositions	Execution Time (ET)	Percentage of Answers (PA)
ANAPSID SSGM decomposition for CD6	36.49	100
D_1	52.50	100
D_2	**9.61**	100
D_3	40.66	100

that group triple patterns that share one variable; each subgoal is associated with the endpoints where the two triple patterns in the subgoal can be executed, i.e., five out of seven relevant endpoints will be contacted in total, i.e., the Number of Contacts to Endpoints (NCE) is 5. Additionally, endpoints that are not in the intersection of the relevant endpoints of the triple patterns assigned to each subquery are not selected, i.e., one endpoint is not selected per triple patterns **1**, **2**, and **3**; thus, the Number of Non-Selected Endpoints (NNME) is 3. Reducing the number of sources can have a great impact on the query execution time but may negatively affect the answer completeness.

We ran CD6 and decompositions D_1, D_2, and D_3 on ANAPSID to evaluate the impact of these decompositions on execution time and number of query results. The studied metrics are *Execution Time (ET)* and *Percentage of Answers (PA)*. *ET* corresponds to the elapsed time measured as the absolute wall-clock system time reported by the Python `time.time()` function, while *PA* reports on the percentage of the answers produced by the engine; duplicates are not considered. CD6 was executed against ten Virtuoso endpoints that locally access the FedBench collections. Table 2 describes the distribution of the FedBench collections per endpoint. This study was executed on a Linux Mint machine with an Intel Pentium Core 2 Duo E7500 2.93 GHz 8 GB RAM 1333 MHz DDR3.

Table 3 reports on *ET* and *PA*. As expected, values of execution time are diverse and are considerably impacted by the number of subqueries and contacted endpoints. Nevertheless, it is important to highlight the reduction on the execution time that is observed when decomposition D_2 is run on ANAPSID. We formalize the problem of source selection and query decomposition in the context of federations of SPARQL endpoints, and provide an approach able to identify decompositions as D_2 that can be efficiently executed without affecting the percentage of answers. Building on existing results from graph theory, we map the SPARQL federated query decomposition problem into the Vertex Coloring Problem such that colorings correspond to decompositions of a SPARQL query against a federation of SPARQL endpoints, and extend an approximate solution of the Vertex Coloring Problem for decomposing SPARQL queries into subqueries against SPARQL endpoints in the federation.

4   *Fed-DSATUR:* An Adaptive Approach to the Problem of Federated Query Processing

In this section we introduce the Federated SPARQL Query Decomposition Problem (FSQD), and an approximate algorithm named *Fed-DSATUR* that exploits properties of Graph theory to provide an efficient and effective solution to FSQD. We first define the FSQD problem, then we briefly describe the Vertex Coloring problem and the mapping of the FSQD problem into the Vertex Coloring problem. Finally, properties of *Fed-DSATUR* are presented.

4.1   The Federated SPARQL Query Decomposition Problem

A federation is a triple $FE = (En, ds, ins)$, where En is a set of SPARQL endpoints. The function $ds(.)$ maps a resource p in I (resources' URIs according to terminology from [21]) to the set of endpoints in En that can answer the triple pattern $\{?s \; p \; ?o\}$, i.e., the endpoints where the query ASK $\{?s \; p \; ?o\}$ evaluates TRUE. Further, $ins(.)$ is a mapping from endpoints in En to RDF graphs, and $ins(e)$ represents the RDF graphs that are accessible via e. The Federated SPARQL Query Decomposition Problem (FSQD) consists of a federation $FE = (En, ds, ins)$ of SPARQL endpoints, and a SPARQL query Q. FSQD corresponds to the problem of decomposing Q into subqueries that can be executed against endpoints in FE, and *minimize* the execution time and *maximize* query answer completeness.

Definition 1 (SPARQL Query Decomposition). *Given a BGP in a query Q executed against a federation $FE = (En, ds, ins)$. Let $D = (DP, f, g)$ be a decomposition of BGP, where:*

- *$DP = \{SQ_1, \ldots, SQ_m\}$ is a partition of the triple patterns in BGP, where SQ_i is a subquery for $1 \le i \le m$;*
- *$f(.)$ maps SQ_i to the endpoints in En where SQ_i will be executed; and*
- *$g(.)$ maps each triple pattern of BGP to the subquery SQ_i where is assigned to in D.*

Consider query CD6 from our running example. CD6 has one *BGP*, and decomposition D_1 is represented as follows: $D_1 = (DP = \{SQ_1, SQ_2, SQ_3, SQ_4\}, f, g)$:

- $f = \{(SQ_1, \{\text{Jamendo, SWDF, DBpedia, Bibliographic}\}),$
    $(SQ_2, \{\text{LMDB, SWDF, DBpedia, Jamendo}\}),$
    $(SQ_3, \{\text{DBpedia, NYT, Geo}\}),$
    $(SQ_4, \{\text{DBpedia, Geo}\})\}$
- $g = \{(1, SQ_1), (2, SQ_2), (3, SQ_3), (4, SQ_4)\}$

Given a decomposition $D = (DP, f, g)$ of query Q, and a *partition component* (i.e., a subquery of DP) $d \in DP$, the execution of d against the endpoints in $f(d)$, $eval(Q, f(d))$, corresponds to the union of the result sets of executing the triple patterns in d on each endpoint in $f(d)$; filters on triple patterns in d are also executed. For example, in the decomposition D_1 of our running example, the result of executing triple SQ_1 on $f(SQ_1)$={LMDB, SWDF, DBpedia, Jamendo} corresponds to the union of the results of executing triple pattern 1 against each of these endpoints. Similarly, the results of SQ_2, SQ_3, and SQ_4 are built.

The execution of Q according to $D = (DP, f, g)$, $eval(Q,D)$, corresponds to the result set of joining the result sets of $eval(Q,f(d))$, for all d in DP. Thus, the answer of CD6 corresponds to the join of the results of executing SQ_1, SQ_2, SQ_3, and SQ_4. In case the WHERE clause of Q comprises several $BGPs$, i.e., $(P_i \ OP \ P_j)$, and OP is UNION or OPTIONAL, then $D_Q = \{D_i\} \cup \{D_j\}$ and $eval(Q,D_Q) = (eval(Q,D_i) \ OP \ eval(Q,D_j))$ where D_i and D_j represent the decompositions of P_i and P_j, respectively. Finally, given a federation $FE = (En, ds, ins)$, $IVF(FE)$ is an integrated view of FE and corresponds to the union of $ins(e)$ for all the e in En; $eval(Q,IVF(FE))$ is the result set of executing Q in $IVF(FE)$. A decomposition D is valid iff for all the federations FE, $eval(Q,D) \subseteq eval(Q,IVF(FE))$.

In order to solve the FSQD, we estimate the cost of a decomposition $D = (DP, f, g)$ of query Q, $cost(D)$, as a trade-off between query execution performance and query completeness. Relying on results that support the efficiency of executing star-shaped queries against SPARQL engines [17,29,31], we assume that queries decomposed with a minimal number star-shaped subqueries will reduce execution time. Further, queries whose evaluation contact the majority of the federation endpoints will result in more complete answers. Thus, $cost(D)$ is defined as the arithmetic mean of the Estimated Triple Pattern Coverage (ESTC), Estimated Number of Non-Selected Endpoints (ENNSE), and Estimated Number of Endpoint Contacts (ENCE). Values of $cost(.)$ belong to [0.0; 1.0]. A value close to 0.0 indicates that the query is decomposed into few subqueries that correspond all to maximal star-shape subqueries, a small number of endpoints are not selected, and the same endpoint is not contacted several times. Contrary, a value close to 1.0 denotes a costly decomposition that meet at least one of the following conditions: (i) Subqueries comprise one triple pattern; (ii) A large number of endpoints are not selected for the evaluation of a triple pattern; and (iii) The same endpoints are contacted several times.

We formally define $cost(.)$ as follows; then we provide an example that illustrates the values of $cost(.)$ for the decompositions of our running example.

Definition 2 (Cost of Query Decomposition). *Given a BGP P of N triple patterns in a query Q executed against a federation $FE = (En, ds, ins)$. Let $D = (DP = \{SQ_1, \ldots, SQ_m\}, f, g)$ be a query decomposition of P, the cost of D, cost(D), is the arithmetic mean of ESTC, ENNSE, and ENCE, where:*

- *ESTC estimates the coverage of the SPARQL triple patterns, and corresponds to the ratio of the number of subqueries in D to the number of triple patterns in P, i.e., $ESTC = \frac{m}{N}$;*
- *ENNSE estimates the number of non-selected endpoints (NNSE) per SPARQL triple pattern[19], i.e., $ENNSE = \sum_{t \in P} \frac{|ds(pred(t)) - (ds(pred(t)) \cap f(g(t)))|}{|ds(pred(t))|}$*

---

[19] $pred(t)$ returns the predicate of the triple pattern t.

– *ENCE estimates the proportion of endpoints that are contacted several times. ENCE is computed as the ratio of the number of contacts to endpoints (NCE) to the number of relevant endpoints, i.e.,* $ENCE = \frac{\sum_{SQ_i \in DP} |f(SQ_i)|}{|\bigcup_{t \in BGP} ds(pred(t))|}$.

To illustrate the proposed cost function, lets first consider a decomposition D_0 of CD6 with only one subquery SQ_1, i.e., SQ_1 comprises all the triple patterns. SQ_1 is associated with DBpedia, the only endpoint where all the triple patterns can be executed. Thus, $cost(D_0)$ is *0.91* and it is computed as the arithmetic mean of: *(i)* ESTC $= \frac{1}{4}$; *(ii)* ENNSE $= (\frac{3}{4} + \frac{3}{4} + \frac{2}{3} + \frac{1}{2})$; *(iii)* ENCE $= \frac{1}{7}$. Table 4 reports on the costs of decompositions D_1, D_2, and D_3. As can be observed, high values of $cost(.)$ suggest that the same endpoints are contacted several times or endpoints required to produce a complete answer are not included in the decomposition. Contacting the same endpoint several times may negatively impact on the execution time, while removing endpoints from the decomposition may conduct to incomplete answers. According to the reported values of $cost(.)$ for decompositions D_1, D_2, and D_3, the best trade off between execution time and answer completeness can be reached evaluating decomposition D_2, where five out of seven endpoints are contacted and only two subqueries of triple patterns that share one variable are executed. These estimates corroborate the results reported in Table 3 where D_2 allows the ANAPSID query engine to reduce execution time, while the percentage of answers remains 100 %.

Table 4. Running example: computing the $cost(.)$ of decompositions. Decompositions of SPARQL query CD6. AVG stands for the arithmetic mean of $cost(.)$ of each decomposition D_1, D_2, and D_3

Decomposition	$cost(.)$
D_1	AVG$(\frac{4}{4}, (\frac{0}{4} + \frac{0}{4} + \frac{0}{3} + \frac{0}{2}), \frac{13}{7}) = 0.95$
D_2	AVG$(\frac{2}{4}, (\frac{1}{4} + \frac{1}{4} + \frac{1}{3} + \frac{0}{2}), \frac{5}{7}) = 0.68$
D_3	AVG$(\frac{3}{4}, (\frac{0}{4} + \frac{0}{4} + \frac{1}{3} + \frac{0}{2}), \frac{10}{7}) = 0.83$

Definition 3 (The Federated SPARQL Query Decomposition Problem (FSQD)). *Given a BGP P in a query Q executed against a federation $FE = (En, ds, ins)$, the Federated SPARQL Query Decomposition Problem (FSQD) identifies a decomposition D of Q such that the cost of the decomposition D, $cost(D)$, is minimal.*

4.2 Mapping the Federated SPARQL Query Decomposition Problem to the Vertex Coloring Problem

The Vertex Coloring Problem refers to coloring graph vertices such that adjacent vertices are colored with different colors and the number of colors is minimized.

Table 5. Notation summary

Q	SPARQL query
U	Infinite set of URIs
B	Infinite set of blank nodes
L	Infinite set of literals
I	Finite set of URIs
$FE = (En, ds, ins)$	Federation of SPARQL endpoints
En	Set of endpoints in the federation FE
$ds(p)$	Mapping from a predicate p to the set of endpoints that access triples *(?s, p, ?o)*
$ins(e)$	RDF graphs accessible through endpoint e
$VCG = (V, E)$	Undirected coloring graph
SC	Set of colors
c	A mapping from set of vertex V to the set of colors SC
$UsedColors(VCG)$	Colors used in a coloring of VCG
cl	A class color, i.e., a color in $UsedColors(VCG)$
BGP	Basic graph pattern, i.e., set of triple patterns
$D = (DP, f, g)$	Query decomposition of BGP
DP	A partition of BGP into subqueries
f	Mapping from subqueries in DP to set of endpoints in En
g	Mapping from triple patterns BGP to the SQ in DP to which is assigned to in D
$eval(Q, f(d))$	Result set of evaluating the conjunction of triple patterns from d in $f(d)$
$eval(Q, D)$	Result set of joining $eval(Q,f(d))$ for all d in DP
$IVF(FE)$	Integrated view of RDF datasets accessible via endpoints in FE
$eval(Q, IVF(FE))$	Result set of evaluating Q in $IVF(FE)$
$cost(D)$	Cost of a decomposition D
$pred(t)$	Predicate of the triple pattern $t = (s, p, o)$
θ	Bijective map from basic graph pattern to vertices in V of a VCG
$var(t)$	Set of variables in triple pattern t

Definition 4 (Vertex Coloring Problem [4]). *Let $VCG = (V, E)$ be a graph named vertex coloring graph. Let c be a mapping from V to SC, where SC is a set of colors. The Vertex Coloring Problem for VCG is to identify for each pair of vertices u and w in V, the values of c, such that the number of colors used in the coloring of the graph is minimized, and $c(u) \neq c(w)$ if there exists an edge*

between u and w in E. UsedColors(VCG) is the subset of SC that corresponds to the set of colors used in the coloring of VCG.

Building on existing results from graph theory, we map the Federated SPARQL Query Decomposition Problem (FSQD) into the Vertex Coloring Problem such that colorings correspond to decompositions of a SPARQL query against a federation of SPARQL endpoints, and extend an approximate solution of the Vertex Coloring Problem to solve this query decomposition problem. We focus on the decomposition of basic graph patterns *BGP*s. Next, we present the definition of the mapping of the Federated SPARQL Query Decomposition Problem (FSQD) into the Vertex Coloring Problem.

Definition 5 (Mapping of FSQD to the Vertex Coloring Problem). *Let P be a basic graph pattern in a SPARQL query Q against a federation $FE = (En, ds, ins)$. Let $VCG = (V, E)$ be a vertex coloring graph built from P as follows:*

- *P and V are homomorphically equivalent, i.e., there is a bijective map θ: $P \rightarrow V$, such that, for each triple pattern t_i in P, there is a node $\theta(t_i)$ in V.*
- *Given two triple patterns t_i and t_j in P, there is an edge $(\theta(t_i), \theta(t_j)) \in E$ if and only if:*
  - *No endpoint in En answers t_i and t_j, i.e., $ds(pred(t_i)) \cap ds(pred(t_j)) = \emptyset$, or*
  - *t_i and t_j in P do not share a variable, i.e., $var(t_i) \cap var(t_j) = \emptyset$.*

Let $D = (DP = \{SQ_1, \ldots, SQ_m\}, f, g)$ be a query decomposition of BGP P in $FE = (En, ds, ins)$. Let c be a mapping from V to SC, where SC is a set of colors, and two vertices share the same color if they are in the same partition component SQ of DP. The Vertex Coloring Problem for the coloring graph VCG is to identify the values of c, such that the number of colors used in the coloring C of the graph is minimized. Given the set UsedColors(VCG) of the colors in SC used in C, number of colors corresponds to the cardinality of DP such that the cost of the decomposition D, cost(D), is minimal.

The mapping of the FSQD Problem to the Vertex Coloring Problem is defined just for one *BGP* of a query *Q*. In case of queries with more than one *BGP*, this correspondence is performed for each *BGP*. Figure 4 illustrates the vertex coloring graph *VCG* created from CD6. *VCG* comprises four nodes, one per triple pattern in CD6; edges represent that the corresponding triple patterns do not share a variable. This restriction will allow for only selecting the same color to nodes that correspond to triple patterns that all are connected by one variable. We rely thus on results reported in the literature which suggest that subqueries of triple patterns that share one variable reduce cardinality of the intermediate results, and may improve the overall execution time of the query [20, 29]. Furthermore, because *VCG* for CD6 is a bipartite graph, and based on the results stated at Preposition 1, an optimal coloring of *VCG* for CD6 can be found, i.e., a decomposition of CD6 where the cost is minimal. Table 5 summarizes the notation used in our proposed formalization.

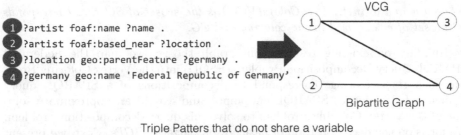

Fig. 4. **Running example: mapping from CD6 to a Vertex Coloring Graph (VCG).** Following Definition 5, nodes in VCG represent triple patterns in CD6, i.e., VCG has four nodes. There is an edge e between nodes n_i and n_j if and only if: (i) Corresponding triple patterns do not share a variable, or (ii) There is no common endpoint that can execute the corresponding triple patterns, i.e., VCG has three edges

4.3   *Fed-DSATUR* a Greedy Algorithm to the Federated SPARQL Query Decomposition Problem

The Vertex Coloring Problem has been shown to be NP-complete [4] and approximate algorithms such as DSATUR, have been defined to solve tractable instances of the problem [28]. DSATUR [4] implements a greedy iterative algorithm that colors each vertex of the graph once by following a heuristic to choose the colors. Given a graph $VCG = (V, E)$, DSATUR orders the vertices in V dynamically based on the number of different colors assigned to the adjacent vertices of each vertex in V, i.e., vertices are chosen based on the degree of saturation on the partial coloring of the graph built so far; only adjacent vertices that are already colored are considered. Intuitively, selecting a vertex with the *maximum degree of saturation* allows to color first those vertices with more restrictions and for which there are smaller sets of colors available. Ties are broken based on the maximum *vertex degree* of the tied vertices, i.e., the *number of adjacent nodes* colored or not; time complexity of DSATUR is $O(|V|^3)$. Further, optimality conditions of the proposed algorithms have received much of attention in the last years; properties of the graphs that are hard to color, in terms of time complexity, for every algorithm [15]. Thus, DSATUR optimally colors most of the graphs that are k-colorable, i.e., k is the number of optimal colors, VCG is k-colorable, and $UsedColors(VCG) \le k$. The following propositions enumerate graphs for which DSATUR is optimal; Fig. 5 illustrates these graphs.

Proposition 1 (Optimality Conditions for DSATUR [15]). *Let $VCG = (V, E)$ be a graph, a core of VCG named $VCG' = (V', E')$, is a sub-graph of VCG, i.e., $V' \subseteq V$ and $E' \subseteq E$, and there is no vertex v in V' such that, $degree(v)$ is 1. DSATUR optimally colors VCG if the core of VCG is as follows:*

Algorithm 1. Fed-DSATUR Algorithm

Input: A SPARQL query Q, a BGP P, and $FE=(En,ds,ins)$
Output: A Decomposition $D=($DP $=\{SQ_1,\ldots,SQ_m\},f,g)$

1 **begin**
2 | DP $\leftarrow \emptyset$; **Generate** VCG=(V,E) from P. // Definition 5
3 | U $\leftarrow$ V , DP $\leftarrow \emptyset$, f $\leftarrow \emptyset$, g $\leftarrow \emptyset$
4 | VC $\leftarrow (0,\ldots,0)$ // Vector for indicating the color of each node
5 | **while** U $\neq \emptyset$ **do**
6 | | UsedColors(VCG) $\leftarrow \emptyset$
7 | | Choose a vertex $v \in$ U with a maximal saturation degree. If there is a tie choose the vertex $v \in$ U of maximum degree
8 | | **foreach** *color* c *used in* VC **do**
9 | | | nodesSameColor $\leftarrow$ obtain vertex set using color c
10 | | | **if** SQ$_i$ *be the subquery corresponding to color* c **then**
11 | | | | f(SQ$_i$) $\leftarrow$ the set of all the triple patterns t where $\theta(t)$ is colored with c
12 | | | | **foreach** $t' \in$ SQ$_i$ **do**
13 | | | | | g(t') $\leftarrow$ SQ$_i$
14 | | | UsedColors(VCG) $\leftarrow$ UsedColors(VCG) $\cup$ {nodesSameColor}
15 | | newColor $\leftarrow$ Choose a subquery SQ$_j$ where the triple pattern t in P that corresponds to vertex v, i.e., $\theta(t) = v$, can be assigned, and the cost of D is minimized, // Definition 5
16 | | **if** c$_j$ *is the legal color that corresponds to* SQ$_j$ **then**
   | | | // ASK SPARQL queries are performed to determine the SPARQL endpoints that can answer SQ$_j$ $\cup$ {t}
17 | | | VC[v] $\leftarrow$ c$_j$ // assign legal color to vertex v
18 | | | U $\leftarrow$ U $\setminus \{v\}$
19 | | | P $\leftarrow$ P $\setminus \{t\}$
20 | | | DP $\leftarrow$ DP $\setminus$ SQ$_j$
21 | | | SQ$_j$ $\leftarrow$ SQ$_j \cup \{t\}$
22 | | | DP $\leftarrow$ DP $\cup$ SQ$_j$
23 | | | f(SQ$_j$) $\leftarrow \bigcap_{t=\{s\,p\,o\}\in SQ_j} ds(p)$
24 | | | **foreach** $t \in$ SQ$_j$ **do**
25 | | | | g(t) $\leftarrow$ SQ$_j$
26 | **return** D $= ($DP$,$f$,$g$)$

- *A single vertex;*
- *A bipartite graph, i.e., a graph that can be partitioned into two set of vertices such that each vertex in each set is connected to a vertex in the other set;*
- *A wheel, i.e., a graph formed by connecting one vertex with all the vertices of a cycle;*
- *A complete multipartite graph, i.e., a graph in which vertices are adjacent if and only if they belong to different partitions of the graph;*

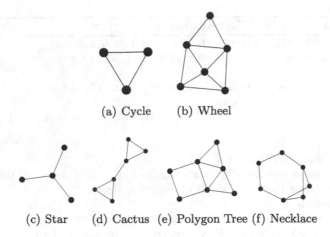

(a) Cycle   (b) Wheel

(c) Star   (d) Cactus (e) Polygon Tree (f) Necklace

Fig. 5. Characterization of graphs for optimal coloring. Graphs where DSATUR produces optimal colorings in polynomial time, i.e., they have a core graph that meets Proposition 1 or are a polygon tree Proposition 2

- *A cactus, i.e., a graph in which any pair of cycles has a vertex in common; and*
- *A necklace, i.e., a graph made up of r beads where each bead is comprised of one cycle of length k which is incident with a path of length l.*

Proposition 2 (Optimality Conditions for DSATUR, Polygon Tree [15]). *DSATUR optimally colors VCG, if VCG is a polygon tree, i.e., (i) VCG is a cycle (Base Case), or (ii) VCG is comprised of two polygon trees VCG' and VCG'' that share exactly one edge.*

As the result of casting FSQD to the Vertex Coloring Problem, we extended the DSATUR algorithm to identify query decompositions where the cost is minimal; we name this extension *Fed-DSATUR*. *Fed-DSATUR* iteratively adds triple patterns to a subquery following the DSATUR heuristic, i.e., the triple patterns that are chosen first, do not share a variable or an endpoint with the *largest* number of triple patterns that have been already assigned to a subquery. Ties between two triple patterns are broken based on the number of triple patterns that do not share a variable or an endpoint with these two triple patterns. The selected triple patterns and their assignments to subqueries ensure that the cost of the decomposition is minimized. A sketch of *Fed-DSATUR* is presented in Algorithm 1; properties of *Fed-DSATUR* are stated in Theorems 1 and 2.

Figure 6 shows the first three iterations of *Fed-DSATUR* conducted during the decomposition of CD6, i.e., three iterations of the **while loop** between lines **5** and **25** of Algorithm 1. Nodes **1** and **4** have the highest *vertex degree* (line **7** of Algorithm 1), i.e., the corresponding triple patterns are the ones with the *largest* number of triple patterns that do not share a variable or an endpoint with them. Thus, *Fed-DSATUR* can choose any of them, and selects node **4** in the iteration

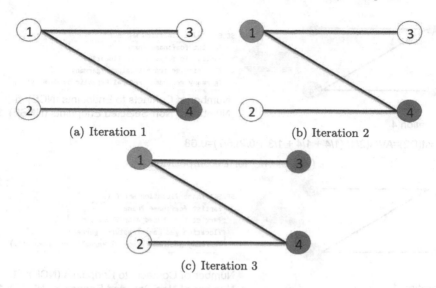

(a) Iteration 1   (b) Iteration 2

(c) Iteration 3

Fig. 6. Running example: iterations of _Fed-DSATUR_. The first three iterations of _Fed-DSATUR_ for CD6

1 (Fig. 6(a)). Then, node **1** is selected in iteration 2 of the **while loop** as shown in Fig. 6(b). The SPARQL endpoints in the federation are contacted using ASK SPARQL queries to determine the ones that can answer the corresponding triple patterns (line **16** of Algorithm 1). Because both nodes are adjacent, they need to be colored with different colors. Nodes **2** and **3** can be chosen in iteration 3 of the **while loop**, and _Fed-DSATUR_ considers all the combinations of colors of each of these two nodes independently. For each coloring, values of the function _cost(.)_ are computed, and the configuration with the lowest value of _cost_ is the one where node **3** is colored in the same color than node **4** (line **15** of Algorithm 1), i.e., triple **3** is placed in the same subquery that triple **4**; Fig. 6(c) reports on the result of this coloring. SPARQL endpoints are contacted to verify that the subquery composed of triple patterns **3** and **4** can be executed in at least one endpoint, i.e., ASK SPARQL queries are posed against the SPARQL endpoints of the federation (line **16** of Algorithm 1).

Finally, in the last iteration of algorithm, node **2** is considered. _Fed-DSATUR_ analyzes if node **2** will be colored with the same color than node **1** or a new color will be used. Once again, the SPARQL endpoints are contacted to verify that the subquery composed of triple patterns **1** and **2** can be executed on the SPARQL endpoints (line **16** in Algorithm 1). Figure 7(a), (b) present the two colorings of the _VCG_ that are considered in Iteration 4 as well as the corresponding decompositions D_2 and D_3; these two decompositions are generated at lines **16–25** of Algorithm 1. Because assigning the same color to nodes **1** and **2** has lower cost than using a new color for coloring node **2**, _Fed-DSATUR_ decides to produce the coloring in Fig. 7(a). At this point, all the triple patterns in CD6 have

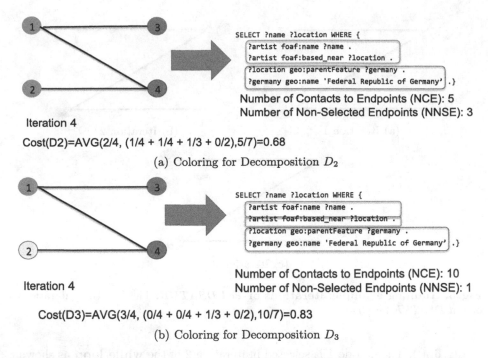

SELECT ?name ?location WHERE {
  ?artist foaf:name ?name .
  ?artist foaf:based_near ?location .
  ?location geo:parentFeature ?germany .
  ?germany geo:name 'Federal Republic of Germany' .}

Number of Contacts to Endpoints (NCE): 5
Number of Non-Selected Endpoints (NNSE): 3

Iteration 4

Cost(D2)=AVG(2/4, (1/4 + 1/4 + 1/3 + 0/2),5/7)=0.68

(a) Coloring for Decomposition D_2

SELECT ?name ?location WHERE {
  ?artist foaf:name ?name .
  ?artist foaf:based_near ?location .
  ?location geo:parentFeature ?germany .
  ?germany geo:name 'Federal Republic of Germany' .}

Number of Contacts to Endpoints (NCE): 10
Number of Non-Selected Endpoints (NNSE): 1

Iteration 4

Cost(D3)=AVG(3/4, (0/4 + 0/4 + 1/3 + 0/2),10/7)=0.83

(b) Coloring for Decomposition D_3

Fig. 7. Running example: iterations of *Fed-DSATUR*. Decompositions considered during Iteration 4 for CD6: (a) decomposition D_2 which is the one the lowest cost; (b) decomposition D_3. Although both decompositions are explored *Fed-DSATUR* only consider D_2 because it is the one that minimizes the values of the function *cost(.)*

been considered, and *Fed-DSATUR* outputs the decomposition D_2. As shown in Table 4, D_2 is the decomposition with the lowest value of the function *cost(.)*.

4.4   Complexity and Formal Proofs

Theorem 1 (FSQD Time Complexity). *The FSQD Problem is NP-Hard.*

Proof 1. *We will reduce a well-known NP-Complete Problem to the FSQD Problem. Let $VCG = (V, E)$ be a graph instance of the Vertex Coloring Problem, we can obtain the equivalent instance of the FSQD Problem given by Definition 5, solve the FSQD problem, and from the output $D = (DP, f, g)$ obtain the solution to the Vertex Coloring Problem as follows: For each partition component d in DP, $\{\theta(t)| \ t \in d \}$ is the set of vertices in V that are colored in color d, and the number of used colors is $cost(D) = k$; we call this solution k-coloring. To conclude the proof, it is enough to show that k is the minimal number of colors needed to color VCG. Suppose there exists, $j < k$ such that, j is the minimal number of colors needed to color VCG and j-coloring a coloring of VCG using j colors, then there exists at least one color ck in k-coloring that is not in j-coloring. Suppose the color ck is used in k-coloring to color the vertex v and let cj be the color* *in j-coloring used to color this vertex v. Because $j < k$, there exists a partition component $d \in DP$ such that, $d = cj$. However, v and $\theta(t)$ for $t \in d$ share the same color in j-coloring and by definition of the Vertex Coloring Problem, there cannot be an edge between them. According to Definition 5, this means that they share a variable and can be evaluated by the same endpoint. Thus, it is false that a solution to the FSQD problem assigns them to different partitions because it would violate the minimality of the cost of the decomposition D.* ◇

Next we rely on the properties of the DSATUR greedy algorithm to state a sufficient condition to be met by a decomposition D to be optimal.

Theorem 2 (Optimality of FSQD). *Let P be a basic graph pattern BGP in a SPARQL query Q. Let $D = (DP, f, g)$ be a query decomposition of P in $FE = (En, ds, ins)$ produced by Fed-DSATUR. Let VCG be the vertex coloring graph obtained from P following Definition 5. If the function $cost(D)$ is monotonic w.r.t. the number of subgoals in DP[20], VCG meets Propositions 1 or 2, and VCG is optimally colored with $UsedColors(VCG) = k$ colors, then DP has k subgoals and $D = (DP, f, g)$ is an optimal decomposition of Q, i.e., there is no other decomposition $D' = (DP', f', g')$, such that: $cost(D') < cost(D)$.*

Proof 2. *Assume that $D = (DP, f, g)$ is not optimal and $cost(D')=j$. Following Definition 5, there is another j-coloring of VCG such that $j \leq UsedColors(VCG)$; nevertheless, by Propositions 1 and 2, k-coloring of VCG is optimal, i.e., VCG is k-colorable and $UsedColors(VCG) \leq k$. Because the function cost is monotonic w.r.t. cardinality of DP, if k is the optimal number of colors, there is no other decomposition $D' = (DP', f', g')$ such that $\mid DP \mid < \mid DP' \mid$ and $cost(D) > cost(D')$, i.e., D is optimal.* ◇

As a result of Theorem 2, if a query comprises only triple patterns that can be executed against only one SPARQL endpoint and the corresponding Vertex Coloring Graph is of the forms stated in Propositions 1 or 2, the decomposition produced by Fed-DSATUR is optimal in terms of the number of star-shaped subqueries and number of query answers.

5 Experimental Study

Datasets and SPARQL Endpoints: We set up two federations of SPARQL endpoints Fed_1 and Fed_2 to access the FedBench collections[21]: DBpedia, NY Times, Geonames, KEGG, ChEBi, Drugbank, Jamendo, LinkedMDB, SW Dog Food, and SP^2B-10M. Both Fed_1 and Fed_2 comprise Virtuoso SPARQL endpoints, where Virtuoso timeout was set up to 240 s. or 71,000 tuples to simulate

---

[20] $cost(D)$ is monotonic w.r.t. number of subgoals in DP, if and only if, the values of $cost(D)$ monotonically increase with the number of subgoals in DP, i.e., for all $D = (DP, f, g)$ and $D' = (DP', f', g')$ such that $\mid DP \mid < \mid DP' \mid$ one has $cost(D) < cost(D')$. A sufficient condition for the function $cost(.)$ to be monotonic is that the query comprises only triple pattern that can be evaluated against only one endpoint.

[21] http://iwb.fluidops.com:7879/resource/Datasets, November 2011.

Table 6. Datasets and SPARQL endpoints. RDF predicates distributed per SPARQL endpoint in federations Fed_1 and Fed_2

Predicate	Fed_1 #Endpoints	Fed_2 #Endpoints
http://www.w3.org/2002/07/owl#sameAs	17	6
http://www.w3.org/1999/02/22-rdf-syntax-ns#type	23	9
http://purl.org/dc/terms/title	2	2
http://xmlns.com/foaf/0.1/name	5	4
http://xmlns.com/foaf/0.1/based_near	12	3
http://www.geonames.org/ontology#parentFeature	12	3
http://www.geonames.org/ontology#name	12	3
http://www.w3.org/2000/01/rdf-schema#label	7	6
http://www.geonames.org/ontology#officialName	11	1
http://xmlns.com/foaf/0.1/based_near	3	3
http://www.w3.org/2004/02/skos/core#subject	2	2
http://bio2rdf.org/ns/bio2rdf#url	2	2
http://purl.org/dc/elements/1.1/title	5	5
http://bio2rdf.org/ns/bio2rdf#xRef	2	2
http://purl.org/dc/elements/1.1/title	5	5

configurations of real-world endpoints [2]; all the endpoints were installed in the same machine. Fed_1 comprises 26 endpoints; 11 of these endpoints access data of *horizontal* and *vertical* fragments of Geonames, and seven SPARQL endpoints access seven DBpedia datasets which correspond to *vertical* fragments of the original collection. Horizontal fragments contain subsets of the triples of an RDF dataset independent of the fact that they share or not the same predicate; description of Fed_1 is presented in Table 1. On the other hand, vertical fragmentation produces fragments which are comprised of all the triples sharing one predicate in the RDF dataset. Horizontal fragmentation impacts on the query answer completeness while vertical fragmentation affects mostly performance. Federation Fed_2 comprises 10 endpoints, one endpoint per collection in the FedBench benchmark (see Table 2). Furthermore, Table 6 presents the RDF predicates that belong to datasets accessible through more than one SPARQL endpoint. This distribution clearly shows that queries against federation Fed_1 that include triple patterns on the predicates http://www.w3.org/2002/07/owl# sameAs or http://www.w3.org/1999/02/22-rdf-syntax-ns#type will require to contact a large number of endpoints to ensure completeness of the answer. The optimization techniques implemented by *Fed-DSATUR* that produce queries with a reduced number of subqueries and cost will allow for a trade-off between number of contacted endpoints and query answer completeness.

Query Benchmarks: We ran the 25 FedBench queries against the collections [26]: Cross-Domain, Linked Data and Life Sciences, which are accessible via the SPARQL endpoints of Fed_1 and Fed_2. Further, we defined ten additional queries (*Additional Queries*, C1-C10) which comprise a large number of triple patterns, basic graph patterns, and different SPARQL operators. Extended setup evaluates the effects of selectivity of *BGPs*, large number of triple patterns, and number of SPARQL operators. The additional queries are composed of between 6 and 46 triple patterns and can be decomposed into up to 9 subqueries. Details of these queries are presented in Appendix A.

Federated Engines: We ran FedBench queries on the state-of-the-art federated engines: SPLENDID [10], FedX [27], and ANAPSID [1,20]. Details of the configurations of each of the engines are presented at Table 7. SPLENDID was configured to select execution plans in terms of a dynamic programming based heuristic that relies on real estimates of the cardinalities of the RDF datasets. FedX was run in both cold and warm caches. FedX in cold cache does not exploit any information about the endpoints and contact all the endpoints in the federation to verify which of these endpoints can answer the triple patterns of a query, i.e., instances of the function *ds* are created on-the-fly. On the other hand, FedX in warm cache exploits information recorded in previous query executions, to find the endpoints that can answer a particular triple pattern, i.e., instances of the function *ds* are built from previous executions and used during source selection. Finally, ANAPSID was set up to use the Star-Shaped Group Multiple endpoint selection (SSGM) heuristic. Following the SSGM source selection heuristics, ANAPSID decomposes *BGPs* into subqueries that can be executed by at least one SPARQL endpoint; only information about the predicates associated with the RDF datasets accessible via the SPARQL endpoints is considered by ANAPSID to perform the source selection task.

Table 7. Experimental set up. Federation SPARQL query engines

SPLENDID [10]	Version November 2012. Source selection corresponds to source selection: ASK, queryOptimization: DYNAMIC_PROGRAMMING, cardEstimator: TRUE_CARD
FedX [27]	Version 1.1 August 2012. Experiments ran on cold and warm caches. FedX in cold cache does not record any information about the endpoints, i.e., the file cache.db was deleted. Cache was warmed up by running five times the same query and reporting the best time
ANAPSID [1,20]	Version May 2014. Endpoint selection technique SSGM, i.e., queries are decomposed into star subqueries that can be evaluated by at least one endpoint

Evaluation Metrics:

- *Execution Time:* Elapsed time between the submission of the query to an engine and the generation of the query answer; optimization time just considers the time elapsed between the submission of the query and the output of the query physical plan. Execution time corresponds to absolute wall-clock system time as reported by the Python time.time() function. We ran each query ten times and reported the average Execution Time.
- *Throughput:* Number of answers produced per second; this is computed as the ratio of the number of answers to execution time in seconds.
- *Percentage of the Answer (PA):* Query answer completeness percentage; ground truths were computed by running each query against a SPARQL endpoint that access via the SPARQL protocol, an integrated view of the FedBench collections, i.e., *IVF(FE)*.

Implementations: *Fed-DSATUR* was implemented on top of ANAPSID [1] query engine using Python 2.6.5. Message size and execution timeout were 16 KB and 600 s, respectively. SPARQL endpoints are installed in a Linux Mint machine with an Intel Pentium Core 2 Duo E7500 2.93 GHz 8 GB RAM 1333 MHz DDR3. Federated SPARQL engines are installed in Linux Debian 8 machine with an CPU Intel I7 980X 3.3 GHz with 24 GB RAM 1333 MHz DDR3. Machine are dedicated to exclusively execute these experiments.

5.1 Analysis of FedBench and Additional Queries

We analyze the evaluated SPARQL queries based on the characteristics of the vertex coloring graphs generated following Definition 5. Table 8 reports on the characteristics of the 25 FedBench and the ten additional queries. In case a query comprises several *BGPs*, we only report on the *BGP* that induces a more complex graph; #SubGoals corresponds to the number of parts or subqueries in the corresponding decomposition; #Nodes and #Edges represent the nodes and edges in the corresponding coloring vertex graph. Based on the properties stated in Propositions 1 and 2 and Theorem 2, depending on the shape of the vertex coloring graphs, optimality conditions of the coloring (i.e., decompositions) found by *Fed-DSATUR* can be ensured whenever *cost(.)* is monotonic.

First, we can observe that *BGPs* of the FedBench queries can be represented as *Disconnected, Bipartite*, or *Tripartite* graphs in both federations. Therefore, the corresponding *BGPs* of these queries can be decomposed by *Fed-DSATUR* in up to three subqueries. Note that optimality in the context of FSQD refers to the decompositions where the cost is minimal and these decompositions can be found by *Fed-DSATUR* in polynomial time. Additionally, we note that in both Fed_1 and Fed_2, the corresponding graphs of the ten additional queries are multipartite graphs composed of a large number of edges, e.g., C1, C3, and C9. Optimal decompositions can be obtained in polynomial time for some of these queries, i.e., queries that meet Theorem 2. However, since time complexity depends on the size of the graph, query decomposers can be considerably impacted during

Table 8. Characterization of FedBench and additional queries in Fed_1 and Fed_2. Based on the casting of Federated SPARQL Query Decomposition Problem (FSQD) to the Vertex Coloring Problem (Definition 5) queries are characterized. If a query comprises several *BGPs*, only the most complex subgraph is reported. #SubGoals corresponds to the number of parts or subqueries in the corresponding decomposition; #Nodes and #Edges represent the nodes and edges in the corresponding coloring vertex graph. Queries, graphs, and colorings are published at http://scast.github.io/fed-dsatur-decompositions/

Query	Fed_1				Fed_2			
	# Nodes	#Edges	#SubGoals	Shape	# Nodes	#Edges	#SubGoals	Shape
CD1	2	0	1	Disconnected	2	0	1	Disconnected
CD2	3	2	2	Bipartite	3	2	2	Bipartite
CD3	5	8	3	Tripartite	5	3	2	Bipartite
CD4	5	3	3	Tripartite	5	7	3	Tripartite
CD5	4	4	2	Bipartite	4	4	2	Bipartite
CD6	4	4	3	Tripartite	4	4	3	Tripartite
CD7	4	1	2	Bipartite	4	1	2	Bipartite
LD1	3	0	1	Disconnected	3	0	1	Disconnected
LD2	3	0	1	Disconnected	3	0	1	Disconnected
LD3	4	2	2	Bipartite	4	0	2	Disconnected
LD4	5	3	2	Bipartite	5	4	2	Bipartite
LD5	3	2	2	Tripartite	3	0	1	Disconnected
LD6	5	9	3	Tripartite	5	7	3	Tripartite
LD7	2	0	1	Disconnected	2	0	1	Disconnected
LD8	5	5	3	Tripartite	5	4	2	Bipartite
LD9	3	3	3	Tripartite	3	2	2	Bipartite
LD10	3	2	2	Bipartite	3	1	2	Bipartite
LD11	5	7	3	Tripartite	5	0	1	Disconnected
LSD1	1	0	1	Disconnected	1	0	1	Disconnected
LSD2	2	1	2	Bipartite	2	1	2	Bipartite
LSD3	5	4	2	Bipartite	5	4	2	Bipartite
LSD4	7	12	3	Bipartite	7	13	3	Bipartite
LSD5	6	11	3	Tripartite	6	11	3	Tripartite
LSD6	5	6	2	Bipartite	5	6	2	Bipartite
LSD7	4	4	2	Bipartite	4	4	2	Bipartite
C1	16	**83**	7	**7-Partite**	16	**85**	7	**7-Partite**
C2	12	36	4	4-Partite	12	38	4	4-Partite
C3	13	**41**	6	**6-Partite**	13	**42**	6	**6-Partite**
C4	19	79	6	4-Partite	19	79	6	4-Partite
C5	6	3	2	Bipartite	6	3	2	Bipartite
C6	2	1	2	Bipartite	5	4	3	Tripartite
C7	7	14	4	4-Partite	7	15	4	4-Partite
C8	7	14	4	4-Partite	7	14	4	4-Partite
C9	40	**500**	9	**9-Partite**	40	**503**	9	**9-Partite**
C10	4	5	3	Tripartite	4	5	3	Tripartite

the decomposition of these queries. Thus, they constitute a challenge for *Fed-DSATUR* and to existing federated query engines. We refer the reader to http://scast.github.io/fed-dsatur-decompositions/ to check the mappings and colorings as well as the results of our experiments.

We have developed a demo [7] named *Silurian*[22] that illustrates the shapes of the vertex coloring graphs, the available endpoints per triple pattern, and different decompositions when queries are executed against federations *Fed*$_1$ and *Fed*$_2$. Figure 8 presents the vertex coloring graph for the additional query C9 when it is executed against *Fed*$_1$. C9 can be optimally decomposed into nine subqueries all executed against the Drugbank endpoint. Nevertheless, because there are 500 edges among 40 nodes, identifying the optimal coloring in this graph is a challenging problem. We note that *Fed-DSATUR* was able to decompose this query, but the execution engine timed out at 600 s without producing any answer. Similar behavior was observed in the other federated query engines. These results suggest that queries that can be mapped to vertex coloring graphs as the one presented in Fig. 8, are hard to decompose and to execute for existing federated query engines, and represent new challenges to the area.

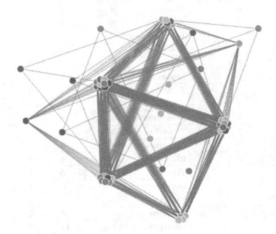

Fig. 8. Vertex Coloring Graph *VCG* **for query C9 on federation** *Fed*$_1$. *VCG* is composed of 40 nodes and 500 edges, and is a 9-partite graph. Nodes represent triple patterns in query C9. There is an edge e between nodes n_i and n_j if and only if: (i) Corresponding triple patterns do not share a variable, or (ii) There is no shared endpoint

5.2   Effectiveness and Efficiency of *Fed-DSATUR* FedBench Queries

The aim of this evaluation is to show the performance of *Fed-DSATUR* and existing engines when FedBench queries are posed against federations *Fed*$_1$ and *Fed*$_2$. Performance of *Fed-DSATUR* and existing engines is measured in terms of *Throughput* and *Percentage of Answer* (PA). Tables 9 and 10 report a pair of *Throughput* (# Answer per seconds) and *Percentage of Answers* (PA).

    In general, we can observe that all the federated engines are competitive and exhibit a similar behavior when the 25 queries of FedBench are executed.

---

[22] http://silurian.thalassa.cbm.usb.ve/.

Table 9. **Effectiveness and efficiency of federated SPARQL engines on Fed-Bench and additional queries in Fed_1.** Entries correspond to pairs of *Throughput* (# Answer per seconds) and *Percentage of Answers* (PA). SPLENDID and FedX (Cold cache) exhibit similar behavior; PA $\simeq$ 100 but lower *Throughput*. ANAPSID, FedX (Warm cache), and Fed-DSATUR have similar behavior; ANAPSID and *Fed-DSATUR* have the highest values of *Throughput* for LSD1: 2,979.05 and 2,689.14, respectively. *Fed-DSATUR* obtained PA $\simeq$ 100 % and high values of *Throughput* in 11 out of 25 FedBench queries. *Fed-DSATUR* also has the highest values of *Throughput* and PA in 8 out of 10 additional queries. The top-2 values of *Throughput* are highlighted in **bold**

Query	Fed_1 (Throughput;PA)				
	ANAPSID	FedX_Cold	FedX_Warm	SPLENDID	Fed-DSATUR
CD1	(17.16;77.77)	(0.81;76.19)	(**71.33**;76.19)	(9.33;65.08)	(**62.67**;77.77)
CD2	(**3.25**;100)	(0.99;100)	(2.08;100)	(0.00;0.00)	(**2.70**;100.00)
CD3	(0.18;40.00)	(**0.51**;40.00)	(**1.65**;40.00)	(0.0;0.00)	(0.10;100.00)
CD4	(0.01;100)	(0.33;100)	(**0.88**;100.00)	(**0.56**;100.00)	(0.17;100.00)
CD5	(3.48;100)	(0.84;100.00)	(2.78;100.00)	(0.00;0.00)	(1.52;100.00)
CD6	(0.0;0.00)	(**1.32**;100.00)	(**3.16**;100.00)	(0.0;0.00)	(0.17;100.00)
CD7	(0.0;0.00)	(0.09;50.00)	(0.18;50.00)	(**0.21**;50.00)	(**1.79**;50.00)
LD1	(**310.91**;100.00)	(23.97;100.00)	(151.86;100.00)	(13.31;98.68)	(**255.08**;100.00)
LD2	(54.39;100.00)	(**174.84**;100.00)	(**344.61**;100.00)	(147.67;100.00)	(153.50;100.00)
LD3	(0.0;0.00)	(5.95;100.00)	(**24.20**;100.00)	(**13.51**;98.16)	(5.73;100.00)
LD4	(115.30;100.00)	(41.13;100.00)	(**99.24**;100.00)	(39.80;98.17)	(**158.79**;100.00)
LD5	(0.0;0.00)	(**0.23**;57.14)	(**22.05**;57.14)	(0.0;100.00)	(1.07;57.14)
LD6	(0.82;50.00)	(**45.01**;50.00)	(**45.27**;50.00)	(0.0;0.00)	(0.0;0.00)
LD7	(0.0;0.00)	(64.09;100.00)	(**117.60**;100.00)	(45.97;100.00)	(**535.41**;45.47)
LD8	(0.0;100.00)	(0.0;100.00)	(0.0;100.00)	(0.0;0.00)	(0.0,100.00)
LD9	(0.0;100.00)	(0.0;100.00)	(0.0;100.00)	(0.0;100.00)	(0.0;100.00)
LD10	(0.0;0.00)	(0.0;0.00)	(0.0;0.00)	(0.0;0.00)	(0.0;0.00)
LD11	(0.0;0.00)	(6.93;47.87)	(**66.94**; 47.87)	(0.0;0.00)	(0.0;0.00)
LSD1	(**2,979.05**;100.00)	(918.42;100.00)	(1,490.84;100.00)	(663.11;100.00)	(**2,689.14**;100.00)
LSD2	(4.51;86.89)	(**40.29**;86.89)	(**279.32**;86.89)	(5.31;85.82)	(18.21;86.89)
LSD3	(**332.66**;100.00)	(121.20;100.00)	(127.26100.00)	(0.0;0.00)	(**311.45**;100.00)
LSD4	(0.38;100.00)	(1.72;100.00)	(**6.30**;100.00)	(**1.75**;100.00)	(0.52;100.00)
LSD5	(20.91;100.00)	(19.82;100.00)	(**39.49**;100.00)	(4.42;100.00)	(**21.52**;100.00)
LSD6	(1.82; 100.00)	(3.92;100.00)	(**23.69**;100.00)	(1.71;100.00)	(**4.45**;100.00)
LSD7	(**81.38**;100.00)	(19.01;8.89)	(24.88;8.89)	(32.20;100.00)	(**103.99**;100.00)
C1	(**99.37**;100.00)	(0.0;0.00)	(0.0;0.00)	(0.0;0.00)	(**90.08**;100.00)
C2	(6.76;100.00)	(0.0;0.00)	(0.0;0.00)	(19.99;100.00)	(**2,306.64**;100.00)
C3	(0.0;0.00)	(0.0;0.00)	(0.0;0.00)	(**2.94**;51.38)	(**1.08**;99.69)
C4	(**1.40**;100.00)	(0.0;0.00)	(0.0;0.00)	(0.0;0.00)	(**7.31**;100.00)
C5	(**922.97**;100.00)	(0.72;0.38)	(2.44;0.38)	(**4,871.06**;100.00)	(758.76;100.00)
C6	(**0.92**;100.00)	(0.0;0.00)	(0.0;0.00)	(0.18;100.00)	(**2.40**;100.00)
C7	(**285.78**;100.00)	(0.0;0.00)	(0.0;0.00)	(0.0;0.00)	(**652.43**;97.80)
C8	(**45.34**;100.00)	(0.0;0.00)	(0.0;0.00)	(0.0;0.00)	(**101.30**;100.00)
C9	(0.0;0.00)	(0.0;0.00)	(0.0;0.00)	(0.0;0.00)	(0.0;0.00)
C10	(0.0;0.00)	(0.0;0.00)	(0.0;0.00)	(0.0;0.00)	(0.0;0.00)

Table 10. Effectiveness and efficiency of federated SPARQL engines on Fed-Bench and additional queries in *Fed₂*. Entries correspond to pairs of *Throughput* (# Answer per seconds) and *Percentage of Answers* (PA). SPLENDID and FedX (Cold cache) exhibit similar behavior; PA $\simeq$ 100 but lower Throughput. ANAPSID, FedX (Warm cache), and *Fed-DSATUR* have similar behavior; ANAPSID and *Fed-DSATUR* have the highest values of Throughput for LSD1: 2,443.22 and 3,199.74, respectively. *Fed-DSATUR* exhibit the highest values of *Throughput* and PA in 8 out of 10 additional queries. The top-2 values of *Throughput* are highlighted in **bold**

Query	*Fed₂* (Throughput;PA)				
	ANAPSID	FedX_Cold	FedX_Warm	SPLENDID	Fed-DSATUR
CD1	(**137.10**;96.83)	(24.96;95.23)	(**108.76**;95.23)	(21.25;96.83)	(17.70;96.83)
CD2	(**11.78**;100.00)	(1.50;100.00)	(**2.19**;100.00)	(0.83;100.00)	(1.65;100.00)
CD3	(**24.24**;60.00)	(3.54;80.00)	(**5.35**;80.00)	(0.94;60.00)	(1.38;60.00)
CD4	(**3.70**;100.00)	(0.67;100.00)	(**1.20**;100.00)	(0.29;100.00)	(0.04;100.00)
CD5	(**3.90**;100.00)	(1.24;100.00)	(**2.24**;100.00)	(0.40;100.00)	(0.45;100.00)
CD6	(0.84;100.00)	(**1.92**;100.00)	(**3.38**;100.00)	(0.36;100.00)	(0.27;100.00)
CD7	(0.0;0.00)	(**0.25**;50.00)	(**0.46**;50.00)	(0.05;50.00)	(0.0.24;50.00)
LD1	(**288.85**;100.00)	(68.44;100.00)	(172.44;100.00)	(20.21;99.68)	(**244.99**;99.68)
LD2	(**1751.08**;100.00)	(262.38;100.00)	(343.92;100.00)	(154.33;100.00)	(**1,000.24**;100.00)
LD3	(4.45;100.00)	(16.46;100.00)	(**39.61**;100.00)	(4.41;98.17)	(**178.74**;98.17)
LD4	(**126.59**;100.00)	(64.15;100.00)	(99.31;100.00)	(40.88;100.00)	(**178.26**;100.00)
LD5	(**52.04**;100.00)	(16.27;100.00)	(**41.70**;100.00)	(8.95;100.00)	(26.38;100.00)
LD6	(15.61;100.00)	(**40.34**;100.00)	(**81.93**;100.00)	(15.71;100.00)	(0.0;0.00)
LD7	(126.60;100.00)	(124.35;100.00)	(**506.54**;100.00)	(55.21;100.00)	(**1,277.34**;100.00)
LD8	(0.0;100.00)	(0.0;100.00)	(0.0;100.00)	(0.0;100.00)	(0.0;100.00)
LD9	(0.0;100.00)	(0.0;100.00)	(0.0;100.00)	(0.0;100.00)	(0.0;100.00)
LD10	(**21.36**;100.00)	(2.32;100.00)	(**3.52**;100.00)	(0.68;100.00)	(1.59;100.00)
LD11	(2.54;100.00)	(**33.21**;100.00)	(**98.12**;100.00)	(9.83;100.00)	(2.91;100.00)
LSD1	(**2,443.22**; 100.00)	(1,125.49;100.00)	(1,492.05;100.00)	(773.13;100.00)	(**3,199.74**;100.00)
LSD2	(**281.35**;87.70)	(189.33;87.70)	(**411.71**;87.70)	(107.93;87.70)	(61.14;87.70)
LSD3	(**681.17**;100.00)	(221.74;100.00)	(252.61;100.00)	(32.91;100.00)	(**335.69**;100.00)
LSD4	(0.40;100.00)	(**2.40**;100.00)	(**6.25**;100.00)	(1.79;100.00)	(0.75;100.00)
LSD5	(**41.45**;100.00)	(22.53;100.00)	(**39.77**;100.00)	(4.53;100.00)	(21.82;100.00)
LSD6	(2.22;100.00)	(**6.05**;100.00)	(**25.94**;100.00)	(1.94;100.00)	(3.24;100.00)
LSD7	(**99.32**;100.00)	(20.37;8.90)	(23.63;8.90)	(**32.74**;100.00)	(10.26;100.00)
C1	(**95.88**;100.00)	(0.0;0.00)	(0.0;0.00)	(0.26;100.00)	(**169.84**;100.00)
C2	(6.33;100.00)	(0.0;0.00)	(0.0;0.00)	(**647.01**;100.00)	(**2,400.53**;100.00)
C3	(0.0;0.00)	(0.0;0.00)	(0.0;0.00)	(**3.19**;51.38)	(**1.07**;98.46)
C4	(**1.20**;100)	(0.0;0.00)	(0.0;0.00)	(0.0;0.00)	(**7.23**;100.00)
C5	(**924.54**;100.00)	(5.40;0.38)	(5.32;0.38)	(148.46;100.00)	(**754.21**;100.00)
C6	(**0.26**;100.00)	(0.0;0.00)	(0.0;0.00)	(0.18;100.00)	(**0.65**;100.00)
C7	(**250.06**;100.00)	(0.0;0.00)	(0.0;0.00)	(0.0;0.00)	(**706.50**;97.80)
C8	(**67.28**;100.00)	(0.0;0.00)	(0.0;0.00)	(0.0;0.00)	(**104.94**;100.00)
C9	(0.0; 0.00)	(0.0;0.00)	(0.0;0.00)	(0.0; 0.00)	(0.0;0.00)
C10	(0.0;0.00)	(0.0;0.00)	(0.0;0.00)	(0.0;0.00)	(0.0;0.00)

All the engines are able to produce a great portion of the answers and exhibit good performance. This similar behavior of all the engines may be caused by the properties that characterize the FedBench queries. As can be observed in Table 8, all the graphs are either disconnected, bipartite or tripartite, and in general low execution cost and highly quality decompositions can be generated, i.e., decompositions with a low number of subqueries where these engines have high *Throughput* and *Percentage of Answer*.

SPLENDID and FedX (Cold cache) exhibit similar behavior in both Fed_1 and Fed_2; PA $\simeq$ 100 but lower *Throughput*; SPLENDID relies on a cost model to estimate cardinality of the results, and exploits this information to find plans that minimize intermediate results. FedX on the other hand, does not consider any cost or estimates, but relies on the Exclusive Group criteria to identify decompositions that benefit answer completeness. ANAPSID, FedX (Warm cache), and *Fed-DSATUR* exhibit better performance in both *Throughput* and *Percentage of Answers* (PA); particularly, ANAPSID and *Fed-DSATUR* have the highest values of *Throughput* for LSD1; in the federation Fed_1 the values are 2,979.05 and 2,689.14, respectively. Further, in the federation Fed_2 ANAPSID and *Fed-DSATUR* have values of throughput for LSD1 of 2,443.22 and 3,199.74, respectively.

It is important to highlight that *Fed-DSATUR* exploits the shape of the input queries, and their corresponding coloring vertex graphs and values of the function $cost(.)$. Thus, *Fed-DSATUR* is able to produce decompositions with an optimal number of subqueries and values of the function $cost(.)$. Based on this property, *Fed-DSATUR* exhibits a slightly better performance than the rest of the engines, particularly in federation Fed_1 where the $cost(.)$ function could more precisely discriminate between costly plans with a smaller number of subqueries from less costly plans with a larger number of subqueries, e.g., CD4 and CD6 are decomposed into three subqueries even the number of optimal subqueries would be two if $cost(.)$ were not considered. Thus, *Fed-DSATUR* produces a percentage of answers PA of 100 in 11 out of 25 queries with high values of throughput in federation Fed_1, and it is only competitive with FedX (Warm cache) which relies on previously stored information about the endpoints to select the relevant sources of a query. To conclude, these results on the 25 FedBench queries suggest that even in queries where existing approaches exhibit a good performance, *Fed-DSATUR* can generate decompositions with low values of $cost(.)$[23] which result in plans that produce large *Percentage of the Answers* and exhibit high values of *Throughput*.

This behavior of *Fed-DSATUR* is more clearly observed when queries with a large number of triple patterns are posed against the federated SPARQL engines. SPLENDID exhibits a relatively good performance, with high values of *Throughput* and *Percentage of Answers* for C2, C3, and C5, and with the maximum *Throughput* in C5 (4,871.06) in federation Fed_1. However, *Fed-DSATUR* query plans are able to: *(i)* Reach high values of PA that ranges from 95 to 100 %, i.e., decompositions are able to produce complete answers; *(ii)* Answer 8 out

---

[23] As indicated in Theorem 2, these decompositions can be optimal depending on the property of monotonicity of the function $cost(.)$.

of 10 queries and speed up the execution time of almost all of the queries; for C2 the execution time is three orders of magnitude lower than ANAPSID and two orders of magnitude lower than SPLENDID. These results suggest that strategies implemented in DSATUR in conjunction with the ones implemented in *Fed-DSATUR* positively impact on the effectiveness of the query decomposition problem solution. Furthermore, properties of DSATUR can be exploited to decide optimality conditions of a query decomposed by *Fed-DSATUR*. This is a unique property of *Fed-DSATUR* that none of the existing federated SPARQL engines is able to ensure.

Finally, it is important to highlight that none of these engines could execute C9 and C10 before timing out at 600 s. We emphasize that both queries induce complex graphs which can have up to 503 edges, SPARQL OPTIONAL operators, or need to be evaluated against a large number of SPARQL endpoints. Thus, they require more than 600 s to be decomposed and executed, and constitute challenges for federated engines and should be included in future benchmarks.

5.3   Efficiency and Effectiveness of the *Fed-DSATUR* Cost-Based Optimization Techniques

This experiment evaluates the effect of the *Fed-DSATUR* cost-based optimization technique. The FedX Exclusive Group technique was implemented on top of the ANAPSID query engine, as well as a version of *Fed-DSATUR* that does not consider the *cost* function, i.e., query plans have the number of subqueries as colors assigned by the original DSATUR algorithm; we call these decomposers Exclusive Group (EG) and Vertex Coloring Graph (CG), respectively. The EG decomposer is a light-weight optimizer that decomposes a query into subqueries that can be exclusively executed by exactly one SPARQL endpoint; it ensures query completeness but execution time can be negatively impacted. The CG decomposer generates plans with the minimal number of subqueries between exact-star groups; CG plans can be efficient at the cost of query completeness. Finally, *Fed-DSATUR* implements a cost model that trades off between answer completeness and query execution; thus, optimization and execution time may be impacted, while a larger number of answers can be generated.

Figures 9 and 10 report on the comparison of the behavior of *Fed-DSATUR* (FD), Exclusive Groups (EG), and Vertex Coloring Graph (CG, *Fed-DSATUR* with no cost function *cost(.)*). **Source Selection**, **Optimization**, **First Tuple**, and **Total** are reported as well as the number of answers produced for each query. As observed, both EG and FD are able to produce the same number of answers when the 25 FedBench queries are executed against federation Fed_1 (Fig. 9), and the planning time (**Source Selection** and **Optimization**) of FD is higher than the time required by CG and EG. However, it is important to notice that once FD identifies a good decomposition, it facilitates the generation of a good execution plan; as a result, FD plans can produce answers faster than EG and CG. Moreover, in case federations do not frequently change, FD execution plans could be computed off-line; thus, FD would provide an effective and efficient solution to the Federated SPARQL Query Decomposition problem.

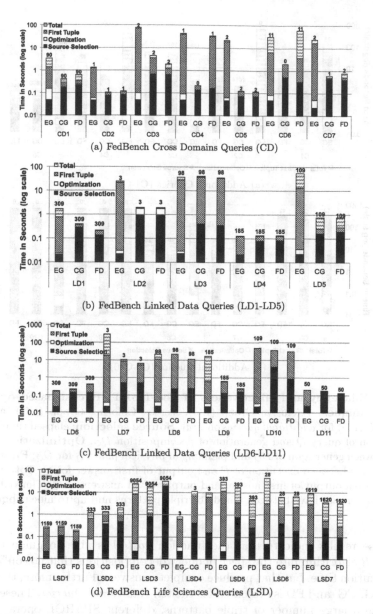

Fig. 9. Efficiency and effectiveness of *Fed-DSATUR* Cost-Based Optimization Techniques. Exclusive Group (EG), Vertex Coloring Graph (CG), and *Fed-DSATUR* (FD) decomposers for FedBench queries on *Fed₁*. **Source Selection:** Elapsed time between the submission of query Q and generation of decomposition D_Q; **Optimization:** Elapsed time between generation of D_Q and planning of physical plan P_Q for D_Q; **First Tuple:** Elapsed time between execution of P and output of first answer for Q; **Total:** Elapsed time between output of first answer and output of all the answers of Q. Bars are annotated with number of answers produced at *Total* time. Timeout is 600 s. (a) Cross Domain (CD); (b, c) Linked Data (LD); (d) Life Sciences (LSD)

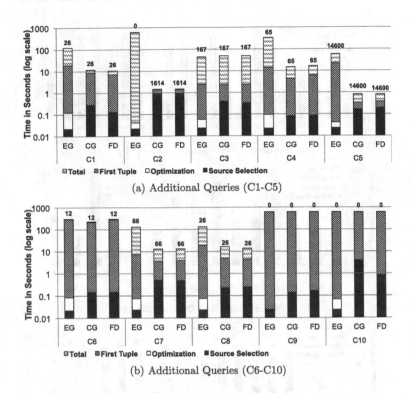

(a) Additional Queries (C1-C5)

(b) Additional Queries (C6-C10)

Fig. 10. Efficiency and effectiveness of Fed-DSATUR cost-based optimization techniques. Exclusive Group (EG), Vertex Coloring Graph (CG), and *Fed-DSATUR* (FD) for *additional queries* on *Fed₁*. **Source selection**: elapsed time between submission of query Q and generation of decomposition D_Q; **Optimization:** elapsed time between generation of D_Q and planning of physical plan P_Q for D_Q; **First Tuple:** elapsed time between execution of P and output of first answer for Q; **Total:** elapsed time between output of first answer and output of all answers of Q. Timeout is 600 s *additional queries* have between 6 and 46 triple patterns and up to nine subqueries

These results indicate that the function *cost(.)* is able to guide the *Fed-DSATUR* algorithm into the space of query decompositions that not only reduce the execution time, but that produce complete answers. Furthermore, the behavior of EG, CG, and FD is evaluated for the ten *Additional Queries*. These queries comprise a large number of triple patterns, different SPARQL operators, and decompositions can be composed of up to nine subqueries (see Table 8); because of these characteristics, these queries represent real challenges for existing federated engines and allow for uncovering hidden properties of federated engines [19]. Figure 10 reports on runtimes and number of answers of these *Additional Queries* when they are executed against *Fed₁*. As observed, *cost(.)* guides FD to find decompositions that not only reduce overall runtime but that are able to produce answers before a timeout of 600 s. These outcomes also suggest that *cost(.)* successfully trades off between query completeness and runtime. Further, although the planning time can be increased, saving achieved during execution time justify the application of the proposed approach.

6   Related Work

The problem of integrating data from dissimilar data sources has been extensively treated in the literature [8,11,12,14,32], and a vast set of integration frameworks [12] have been developed to query heterogeneous sources by implementing the mediator and wrapper architecture proposed by Wiederhold [30]. Furthermore, the Semantic Data Management community has been very active proposing solutions to the problem of query processing on the Web of Data [1,3,10,13,18,23,24,27]. We mainly focus on approaches that implement strategies to address the problem of source selection and decomposition of SPARQL queries, although, we recognize the tremendous advance that the Data Base community has done to the general problem of data integration in the last fifteen years. Existing approaches are grouped according to the amount of knowledge that describes the data sources, and that is exploited during source selection and query decomposition to enhance the quality of the generated query decompositions. First, we describe techniques that rely on index-based structures that encode informations about the datasets and efficiently exploit this information to select the endpoints and RDF documents that can be used to answer a particular query. Next, federated SPARQL query approaches that maintain cardinality and similarity statistics are presented. Finally, we summarize source selection and query decomposition approaches that in presence of zero-knowledge about statistics that describe the cardinality or selectivity of the federated RDF datasets, are able to identify query decompositions against federations of SPARQL endpoints.

6.1   Index-Based Approaches for Source Selection and Query Decomposition

Several approaches have considered index-based approaches to solve the problem of selecting query data providers. Harth et al. [13] present a hybrid solution that combines histograms and R-trees; histograms are used for source ranking, while regions determine the best sources to answer a basic graph pattern. Li and Heflin [18] propose a bottom-up tree based technique to integrate data from multiple heterogeneous sources, and rely on an inverted-index from RDF documents to URIs and literals that identify the set of RDF datasets that can be used to execute a basic graph pattern. Kaoudi et al. [16] propose a P2P system running on top of Atlas for processing distributed RDF documents, and it implements a cost-based approach to identify an optimal plan; statistics are kept by peers. DAW [24] implements duplicate-aware strategies for selecting SPARQL endpoints to execute federated SPARQL queries, in a way that overlapping among the selected endpoints is reduced. DAW relies on Min-Wise Independent Permutations (MIPs) [5], and information about the selectivity of triple patterns in the federation datasets to estimate the overlap among the selected sources. Finally, Saleeem and Ngonga propose the source selection approach named HiBISCuS [23]; HiBISCuS describes RDF data accessible via SPARQL endpoints in terms of capabilities, where a capability states for each predicate p in a dataset the set of URI authorities of the subjects and objects that are related through p. Further, SPARQL queries are represented as labelled hypergraphs to encode in the same hyper-edge, the set of triple patterns that share at least one variable. Capabilities and the hyper-graph based query representation are exploited during source selection and query decomposition to identify the endpoints that will not contribute to the final answer of a query. These approaches can effectively identify data sources for executing a SPARQL query triple patterns. Nevertheless, they all rely on index-based structures build on top of meta-data and statistics about autonomous RDF datasets. Therefore, keeping up-to-date all these indices may be costly in real-world scenarios where data may constantly change or being affected by unexpected environment conditions.

6.2   Cost Based Approaches

DARQ [22] and SPLENDID [10] rely on source descriptions and statistics that describe data accessible via SPARQL endpoints, to perform source selection and query optimization. Both engines decompose queries into subqueries comprised of triple patterns that are uniquely answered by one endpoint or that can be answered by several endpoints. In the later case, triple patterns are sent individually to every relevant endpoint; also, statistics about selectivities and cardinalities are used to identify optimal plans. However, in presence of subqueries associated with more than one endpoint, both systems address the query decomposition problem differently. DARQ uses statistics from endpoint descriptions to build a plan that reduces the number of empty results. Contrary, SPLENDID contacts endpoints to decide if a predicate can be answered. SPARQL-DQP [6] and ARQ[24] exploit information on SPARQL 1.1 queries to decide where the subqueries of triple patterns will be executed. Additionally, they rely on statistics or heuristics to identify query plans. WoDQA[25] is a tool built on top of ARQ to provide access to federations of endpoints. Finally, Avalanche [3] implements an inter-operator solution where a group of best plans is chosen heuristically. Statistics about cardinalities and data distribution are considered to identify possibly good plans. Avalanche follows a competition strategy where top-k plans are executed in parallel, the output of a query corresponds to the answers produced by the most promising plans in a given period of time. These systems are able to achieve good performance and query answer completeness. However, in presence of data sources that constantly change, their behavior can be negatively impacted because estimates can easily become imprecise and inaccurate.

---

[24] http://jena.sourceforge.net/ARQ.
[25] https://github.com/seagent/WoDQA/.

6.3   Adaptive Approaches for Source Selection and Query Decomposition

We describe the approaches that do not resort to statistics, estimates, or index-based structures to decide which endpoints are more suitable to execute a query. FedX [27] and ANAPSID [1] are exemplars of these type of federated engines. Both engines pose ASK and SELECT SPARQL queries against the endpoints to create the instances of the ds function, i.e., to associate a predicate p with the SPARQL endpoints that can answer triple patterns of the form $\{?s\ p\ ?o\}$. FedX [27] is a rule-based system able to decompose SPARQL queries into sub-queries that can be completely executed by an endpoint or *exclusive groups*. FedX uses neither knowledge about mappings nor statistics associated with the sources; it may contact every source to determine where the predicates presented in a query are offered, and may save this information in cache for future queries of the same predicate. FedX relies on the number of bound variables in a query to decide the order in which subqueries will be posed to the selected endpoints. ANAPSID [1] exploits information about the predicates in federation datasets during source selection and query decomposition. ANAPSID identifies the triple patterns that can be executed by the same endpoint and comprise a star. Triple patterns are grouped together in a way that, the number of operations done by one endpoint are maximized while the size of intermediate results and bandwidth are minimized. These federated approaches may efficiently identify a query relevant endpoints. However, because none of these engines implements the source selection and query optimization task as an optimization problem, when queries are comprised of a large number of triple patterns or there are many relevant endpoints, they may produce query decompositions that may perform poorly, i.e., query decompositions that do not minimize execution cost or maximize the answer completeness. *Fed-DSATUR* overcomes this limitation, and provides a greedy based solution that depending on the shape of the SPARQL query is able to produce optimal decompositions.

7   Conclusions and Future Work

We have addressed the problem of source selection and query decomposition of federations of SPARQL endpoints. The formalization of the federated SPARQL query decomposition problem as the Vertex Coloring Problem was devised, and built on results from graph theory, we defined a family of queries for which optimal decompositions can be identified in polynomial time. Furthermore, we proposed the query decomposition algorithm named *Fed-DSATUR* which extends DSATUR to efficiently identify SPARQL query decompositions. We empirically evaluated *Fed-DSATUR* and state-of-the-art federated SPARQL query engines on the FedBench benchmark and in an additional set of *complex* queries. We set up two different federations of SPARQL endpoints to access the data collections of FedBench, and attempted to stress source selection and query decomposition techniques when data collections were fragmented across different datasets accessible via SPARQL endpoints. The observed results suggest that *Fed-DSATUR* and existing SPARQL query engines are competitive in queries comprised of a small number of triple patterns or SPARQL operators. Nevertheless, in presence of *complex* queries composed of a large number of *BGPs* and SPARQL operators, *Fed-DSATUR* can clearly overcome existing approaches in both execution time and answer completeness. In high number of queries, *Fed-DSATUR* could produce almost 100 % of the query answers and exhibit high values of throughput. Additionally, our results pose strong arguments on the type of queries that existing engines are able to execute and when they may fail. In the future, we plan to extend *Fed-DSATUR* to manage *complex* queries such as C9 and C10. Further we will analyze the benefits of considering richest descriptions of the SPARQL endpoints, e.g., range and domain of each RDF predicate, authority fragments of the URL of the RDF datasets, RDF predicates associated with a class in a dataset, and classes in an RDF dataset. We hypothesize that this information can positively impact on the cost function in scenarios with a large number of SPARQL endpoints that share similar RDF data.

A   Additional Complex Queries

We have defined a set of ten additional queries which comprise a large number of triple patterns, basic graph patterns, and different SPARQL operators. Extended setup evaluates the effects of selectivity of *BGPs*, large number of triple patterns, and number of SPARQL operators. The additionally queries are composed of between 6 and 46 triple patterns and can be decomposed into up to 9 subqueries.

Listing 1.2. C1

```
SELECT DISTINCT ?drug ?enzyme ?reaction    Where {
    ?drug1 drugbank:drugCategory drugbankcategory
       : antibiotics .
    ?drug2 drugbank:drugCategory drugbankcategory
       : antiviralAgents .
    ?drug3 drugbank:drugCategory drugbankcategory
       : antihypertensiveAgents .
    ?I1 drugbank:interactionDrug2 ?drug1 .
    ?I1 drugbank:interactionDrug1 ?drug .
    ?I2 drugbank:interactionDrug2 ?drug2 .
    ?I2 drugbank:interactionDrug1 ?drug .
    ?I3 drugbank:interactionDrug2 ?drug3 .
    ?I3 drugbank:interactionDrug1 ?drug .
    ?drug owl:sameAs ?drug5 .
    ?drug5 rdf:type dbpedia:Drug .
    ?drug drugbank:keggCompoundId ?cpd .
    ?enzyme kegg:xSubstrate ?cpd .
    ?enzyme rdf:type kegg:Enzyme .
    ?reaction kegg:xEnzyme ?enzyme .
    ?reaction kegg:equation ?equation .
}
```

Listing 1.3. C2

```
SELECT DISTINCT ?drug ?enzyme ?reaction    Where {
    ?drug1 drugbank:drugCategory drugbankcategory:
       antibiotics .
    ?drug2 drugbank:drugCategory drugbankcategory:
       antiviralAgents .
    ?drug3 drugbank:drugCategory drugbankcategory:
       antihypertensiveAgents .
    ?drug4  drugbank:drugCategory drugbankcategory:anti
       −bacterialAgents .
    ?I1 drugbank:interactionDrug2 ?drug1 .
    ?I1 drugbank:interactionDrug1 ?drug .
    ?I2 drugbank:interactionDrug2 ?drug2 .
    ?I2 drugbank:interactionDrug1 ?drug .
    ?I3 drugbank:interactionDrug2 ?drug3 .
    ?I3 drugbank:interactionDrug1 ?drug .
    ?I4 drugbank:interactionDrug2 ?drug4 .
    ?I4 drugbank:interactionDrug1 ?drug .
    ?drug owl:sameAs ?drug5 .
    ?drug5 rdf:type dbpedia:Drug .
    ?drug drugbank:keggCompoundId ?cpd .
    ?enzyme kegg:xSubstrate ?cpd .
    ?enzyme rdf:type kegg:Enzyme .
    ?reaction kegg:xEnzyme ?enzyme .
    ?reaction kegg:equation ?equation .
}
```

Listing 1.5. C4

```
SELECT DISTINCT ?d ?drug5 ?cpd ?enzyme ?equation
WHERE {
    ?drug1 drugbank:possibleDiseaseTarget diseasome:261.
    ?drug1 drugbank:target ?o.
    ?o drugbank:genbankIdGene ?g.
    ?o drugbank:locus ?l.
    ?o drugbank:molecularWeight ?mw.
    ?o drugbank:hprdId ?hp.
    ?o drugbank:swissprotName ?sn.
    ?o drugbank:proteinSequence ?ps.
    ?o drugbank:generalReference ?gr.
    ?drug drugbank:target?o.
    ?drug owl:sameAs ?drug5 .
    ?drug drugbank:possibleDiseaseTarget ?d1 .
    ?drug owl:sameAs ?drug5 .
    ?drug5 rdf:type dbpedia:Drug .
    ?drug drugbank:keggCompoundId ?cpd .
    ?enzyme kegg:xSubstrate ?cpd .
    ?enzyme rdf:type kegg:Enzyme .
    ?reaction kegg:xEnzyme ?enzyme .
    ?reaction kegg:equation ?equation .
}
```

Listing 1.4. C3

```
SELECT DISTINCT ?drug ?enzyme ?reaction WHERE {
    ?drug1 drugbank:possibleDiseaseTarget diseasome:302.
    ?drug2 drugbank:possibleDiseaseTarget diseasome:53 .
    ?drug3 drugbank:possibleDiseaseTarget diseasome:59 .
    ?drug4 drugbank:possibleDiseaseTarget diseasome:105 .
    ?drug1 drugbank:possibleDiseaseTarget ?d .
    ?drug2 drugbank:possibleDiseaseTarget ?d .
    ?drug3 drugbank:possibleDiseaseTarget ?d .
    ?drug4 drugbank:possibleDiseaseTarget ?d .
    ?drug drugbank:possibleDiseaseTarget ?d .
    ?drug drugbank:casRegistryNumber ?id .
    ?keggDrug rdf:type kegg:Drug .
    ?keggDrug bio2rdf:xRef ?id .
    ?keggDrug dc:title ?title .
}
```

Listing 1.6. C5

```
SELECT DISTINCT ?drug5 ?drug6
WHERE {
    ?drug1 drugbank:possibleDiseaseTarget  diseasome:319 .
    ?drug1 drugbank:possibleDiseaseTarget> diseasome:270 .
    ?l1 drugbank:interactionDrug1 ?drug1 .
    ?l1 drugbank:nteractionDrug2 ?drug .
    ?drug1 owl:sameAs ?drug5 .
    ?drug owl:sameAs ?drug6 .
}
```

Listing 1.7. C6

```
SELECT DISTINCT ?drug
WHERE
{
    ?drug drugbank:drugCategory drugbankcategory:micronutrient .
    ?drug drugbank:drugCategory drugbankcategory:dietarySupplement .
    ?drug drugbank:drugCategory drugbankcategory:non—essentialAminoAcids .
    OPTIONAL {
        ?drug drugbank:indication?i .
        ?drug drugbank:biotransformation ?b .
        ?drug drugbank:inchiKey ?k .
        ?drug drugbank:synonym?s .
        ?drug drugbank:toxicity?t .
    } .
    OPTIONAL {
        ?drug drugbank:keggCompoundId?cpd.
        ?enzyme rdf:type kegg:Enzyme.
        ?enzyme kegg:xSubstrate ?cpd.
        ?reaction kegg:xEnzyme?enzyme.
        ?reaction kegg:equation?equation .
    } .
    OPTIONAL {
        ?drug5 rdf:type dbpedia:Drug .
        ?drug rdf:type dbpedia:Drug
    }
}
```

Listing 1.8. C7

```
SELECT   DISTINCT ?d ?drug5 ?cpd ?enzyme ?equation
WHERE {
        ?drug1 drugbank:possibleDiseaseTarget diseasome:261 .
        ?drug1 drugbank:target ?o .
        ?o drugbank:genbankIdGene ?g .
        ?o drugbank:locus ?l .
        ?o drugbank:molecularWeight ?mw .
        ?o drugbank:hprdId ?hp .
        ?o drugbank:swissprotName ?sn .
        ?o drugbank:proteinSequence ?ps .
        ?o drugbank:generalReference ?gr .
        ?drug drugbank:target ?o .
        OPTIONAL {
            ?drug owl:sameAs ?drug5 .
            ?drug5 rdf:type dbpedia:Drug .
            ?drug drugbank:keggCompoundId ?cpd .
            ?enzyme kegg:xSubstrate ?cpd .
            ?enzyme rdf:type kegg:Enzyme .
            ?reaction kegg:xEnzyme ?enzyme .
            ?reaction kegg:equation ?equation .
        }
}
```

Listing 1.9. C8

```
SELECT  DISTINCT ?drug1
WHERE {
        ?drug1 drugbank:possibleDiseaseTarget <http://www4.wiwiss.fu−berlin.de/diseasome/resource/diseases/673>

        ?drug1 drugbank:target ?o.
        ?o drugbank:genbankIdGene ?g.
        ?o drugbank:locus ?l.
        ?o drugbank:molecularWeight ?mw.
        ?o drugbank:hprdId ?hp.
        ?o drugbank:swissprotName ?sn.
        ?o drugbank:proteinSequence ?ps.
        ?o drugbank:generalReference ?gr.
        ?drug drugbank:target ?o.
        ?drug drugbank:synonym?o1 .
        OPTIONAL {
            ?drug owl:sameAs ?drug5 .
            ?drug5 rdf:type dbpedia:Drug .
            ?drug drugbank:keggCompoundId ?cpd .
            ?enzyme kegg:xSubstrate ?cpd .
            ?enzyme rdf:type kegg:Enzyme .
            ?reaction kegg:xEnzyme ?enzyme .
            ?reaction kegg:equation ?equation .
        }
}
```

Listing 1.10. C9

```
SELECT DISTINCT ?drug ?drug1 ?drug2 ?drug3 ?drug4   ?d1 WHERE {
    ?drug1  drugbank:drugCategory  <http://www4.wiwiss.fu-berlin.de/drugbank/resource/drugcategory/antibiotics
        > .
    ?drug2  drugbank:drugCategory  <http://www4.wiwiss.fu-berlin.de/drugbank/resource/drugcategory/
        antiviralAgents> .
    ?drug3  drugbank:drugCategory  <http://www4.wiwiss.fu-berlin.de/drugbank/resource/drugcategory/
        antihypertensiveAgents> .
    ?drug4  drugbank:drugCategory  <http://www4.wiwiss.fu-berlin.de/drugbank/resource/drugcategory/anti-
        bacterialAgents> .
    ?drug1  drugbank:target ?o1 .
    ?o1  drugbank:genbankIdGene ?g1 .
    ?o1  drugbank:locus ?l1 .
    ?o1  drugbank:molecularWeight ?mw1 .
    ?o1  drugbank:hprdId ?hp1 .
    ?o1  drugbank:swissprotName ?sn1 .
    ?o1  drugbank:proteinSequence ?ps1 .
    ?o1  drugbank:generalReference ?gr1 .
    ?drug  drugbank:target ?o1 .
    ?drug2  drugbank:target ?o2 .
    ?o1  drugbank:genbankIdGene ?g2 .
    ?o2  drugbank:locus ?l2 .
    ?o2  drugbank:molecularWeight ?mw2 .
    ?o2  drugbank:hprdId ?hp2 .
    ?o2  drugbank:swissprotName ?sn2 .
    ?o2  drugbank:proteinSequence ?ps2 .
    ?o2  drugbank:generalReference ?gr2 .
    ?drug  drugbank:target ?o2 .
    ?drug3  drugbank:target ?o3 .
    ?o3  drugbank:genbankIdGene ?g3 .
    ?o3  drugbank:locus ?l3 .
    ?o3  drugbank:molecularWeight ?mw3 .
    ?o3  drugbank:hprdId ?hp3 .
    ?o3  drugbank:swissprotName ?sn3 .
    ?o3  drugbank:proteinSequence ?ps3 .
    ?o3  drugbank:generalReference ?gr3 .
    ?drug  drugbank:target ?o3 .
    ?drug4  drugbank:target ?o4 .
    ?o4  drugbank:genbankIdGene ?g4 .
    ?o4  drugbank:locus ?l4 .
    ?o4  drugbank:molecularWeight ?mw4 .
    ?o4  drugbank:hprdId ?hp4 .
    ?o4  drugbank:swissprotName ?sn4 .
    ?o4  drugbank:proteinSequence ?ps4 .
    ?o4  drugbank:generalReference ?gr4 .
    ?drug  drugbank:target ?o4 .
    OPTIONAL{
        ?l1  drugbank:interactionDrug2 ?drug1 .
        ?l1  drugbank:interactionDrug1 ?drug .
        ?l2  drugbank:interactionDrug2 ?drug2 .
        ?l2  drugbank:interactionDrug1 ?drug .
        ?l3  drugbank:interactionDrug2 ?drug3 .
        ?l3  drugbank:interactionDrug1 ?drug .
        ?l4  drugbank:interactionDrug2 ?drug4 .
        ?l4  drugbank:interactionDrug1 ?drug .
    }
}
```

Listing 1.11. C10

```
SELECT ?title ?actor ?news ?director ?film ?n ?genre WHERE {
    ?film  dc:title 'Tarzan' .
    ?film  lmdb:actor ?actor .
    ?film  lmdb:production_company <http://data.linkedmdb.org/resource/production_company/15> .
    ?actor owl:sameAs ?x .
    OPTIONAL{
        ?x dbpedia:director  ?director .
        ?director dbpedia:nationality ?n .
        ?film lmdb:genre ?genre .
        OPTIONAL {
            ?y owl:sameAs ?x.
        } .
        ?y nyt:topicPage ?news
    } .
}
```

References

1. Acosta, M., Vidal, M.-E., Lampo, T., Castillo, J., Ruckhaus, E.: ANAPSID: an adaptive query processing engine for SPARQL endpoints. In: Aroyo, L., Welty, C., Alani, H., Taylor, J., Bernstein, A., Kagal, L., Noy, N., Blomqvist, E. (eds.) ISWC 2011, Part I. LNCS, vol. 7031, pp. 18–34. Springer, Heidelberg (2011)
2. Buil-Aranda, C., Hogan, A., Umbrich, J., Vandenbussche, P.-Y.: SPARQL web-querying infrastructure: ready for action? In: Alani, H. (ed.) ISWC 2013, Part II. LNCS, vol. 8219, pp. 277–293. Springer, Heidelberg (2013)
3. Basca, C., Bernstein, A.: Querying a messy web of data with Avalanche. J. Web Semant. **26**, 1–28 (2014)
4. Brélaz, D.: New methods to color vertices of a graph. Commun. ACM **22**(4), 251–256 (1979)
5. Broder, A.Z., Charikar, M., Frieze, A.M., Mitzenmacher, M.: Min-wise independent permutations. J. Comput. Syst. Sci. **60**(3), 630–659 (2000)
6. Buil-Aranda, C., Arenas, M., Corcho, O.: Semantics and optimization of the SPARQL 1.1 federation extension. In: Antoniou, G., Grobelnik, M., Simperl, E., Parsia, B., Plexousakis, D., De Leenheer, P., Pan, J. (eds.) ESWC 2011, Part II. LNCS, vol. 6644, pp. 1–15. Springer, Heidelberg (2011)
7. Castillo, S., Palma, G., Vidal, M.: SILURIAN: a SPARQL visualizer for understanding queries and federations. In: Proceedings of the ISWC Posters and Demonstrations Track, pp. 137–140 (2013)
8. Florescu, D., Levy, A.Y., Mendelzon, A.O.: Database techniques for the world-wide web: a survey. SIGMOD Record **27**(3), 59–74 (1998)
9. Fundulaki, I., Auer, S.: Linked open data - introduction to the special theme. ERCIM News **96**, 2014 (2014)
10. Görlitz, O., Staab, S.: SPLENDID: SPARQL endpoint federation exploiting VOID descriptions. In: Proceedings of the International Workshop on Consuming Linked Data (COLD) (2011)
11. Halevy, A.Y.: Answering queries using views: a survey. VLDB J. **10**(4), 270–294 (2001)
12. Halevy, A.Y., Rajaraman, A., Ordille, J.J.: Data integration: the teenage years. In: Proceedings of the 32nd International Conference on Very Large Data Bases (VLDB), pp. 9–16 (2006)
13. Harth, A., Hose, K., Karnstedt, M., Polleres, A., Sattler, K.-U., Umbrich, J.: Data summaries for on-demand queries over linked data. In: Proceedings of the 19th International Conference on World Wide Web (WWW), pp. 411–420 (2010)
14. Ives, Z.G., Halevy, A.Y., Mork, P., Tatarinov, I.: Piazza: mediation and integration infrastructure for semantic web data. J. Web Semant. **1**(2), 155–175 (2004)
15. Janczewski, R., Kubale, M., Manuszewski, K., Piwakowski, K.: The smallest hard-to-color graph for algorithm DSATUR. Discrete Math. **236**(1–3), 151–165 (2001)
16. Kaoudi, Z., Kyzirakos, K., Koubarakis, M.: SPARQL query optimization on top of DHTs. In: Patel-Schneider, P.F. (ed.) ISWC 2010, Part I. LNCS, vol. 6496, pp. 418–435. Springer, Heidelberg (2010)
17. Lampo, T., Vidal, M.-E., Danilow, J., Ruckhaus, E.: To cache or not to cache: the effects of warming cache in complex SPARQL queries. In: Meersman, R., Dillon, T., Herrero, P., Kumar, A., Reichert, M., Qing, L., Ooi, B.-C., Damiani, E., Schmidt, D.C., White, J., Hauswirth, M., Hitzler, P., Mohania, M. (eds.) OTM 2011, Part II. LNCS, vol. 7045, pp. 716–733. Springer, Heidelberg (2011)

18. Li, Y., Heflin, J.: Using reformulation trees to optimize queries over distributed heterogeneous sources. In: Patel-Schneider, P.F. (ed.) ISWC 2010, Part I. LNCS, vol. 6496, pp. 502–517. Springer, Heidelberg (2010)

19. Montoya, G., Vidal, M.-E., Corcho, O., Ruckhaus, E., Buil-Aranda, C.: Benchmarking federated SPARQL query engines: are existing testbeds enough? In: Cudré-Mauroux, P. (ed.) ISWC 2012, Part II. LNCS, vol. 7650, pp. 313–324. Springer, Heidelberg (2012)

20. Montoya, G., Vidal, M.-E., Acosta, M.: A heuristic-based approach for planning federated SPARQL queries. In: Proceedings of the International Workshop on Consuming Linked Data (COLD) (2012)

21. Pérez, J., Arenas, M., Gutierrez, C.: Semantics and complexity of SPARQL. ACM Trans. Database Syst. **34**(3), 16 (2009)

22. Quilitz, B., Leser, U.: Querying distributed RDF data sources with SPARQL. In: Bechhofer, S., Hauswirth, M., Hoffmann, J., Koubarakis, M. (eds.) ESWC 2008. LNCS, vol. 5021, pp. 524–538. Springer, Heidelberg (2008)

23. Saleem, M., Ngonga Ngomo, A.-C.: HiBISCuS: hypergraph-based source selection for SPARQL endpoint federation. In: Presutti, V., d'Amato, C., Gandon, F., d'Aquin, M., Staab, S., Tordai, A. (eds.) ESWC 2014. LNCS, vol. 8465, pp. 176–191. Springer, Heidelberg (2014)

24. Saleem, M., Ngonga Ngomo, A.-C., Xavier Parreira, J., Deus, H.F., Hauswirth, M.: DAW: duplicate-AWare federated query processing over the web of data. In: Alani, H., Kagal, L., Fokoue, A., Groth, P., Biemann, C., Parreira, J.X., Aroyo, L., Noy, N., Welty, C., Janowicz, K. (eds.) ISWC 2013, Part I. LNCS, vol. 8218, pp. 574–590. Springer, Heidelberg (2013)

25. Schmachtenberg, M., Bizer, C., Paulheim, H.: Adoption of the linked data best practices in different topical domains. In: Mika, P. (ed.) ISWC 2014, Part I. LNCS, vol. 8796, pp. 245–260. Springer, Heidelberg (2014)

26. Schmidt, M., Görlitz, O., Haase, P., Ladwig, G., Schwarte, A., Tran, T.: FedBench: a benchmark suite for federated semantic data query processing. In: Aroyo, L. (ed.) ISWC 2011, Part I. LNCS, vol. 7031, pp. 585–600. Springer, Heidelberg (2011)

27. Schwarte, A., Haase, P., Hose, K., Schenkel, R., Schmidt, M.: FedX: optimization techniques for federated query processing on linked data. In: Aroyo, L. (ed.) ISWC 2011, Part I. LNCS, vol. 7031, pp. 601–616. Springer, Heidelberg (2011)

28. Segundo, P.S.: A new DSATUR-based algorithm for exact vertex coloring. Comput. Oper. **39**(7), 1724–1733 (2012)

29. Vidal, M.-E., Ruckhaus, E., Lampo, T., Martínez, A., Sierra, J., Polleres, A.: Efficiently joining group patterns in SPARQL queries. In: Aroyo, L., Antoniou, G., Hyvönen, E., ten Teije, A., Stuckenschmidt, H., Cabral, L., Tudorache, T. (eds.) ESWC 2010, Part I. LNCS, vol. 6088, pp. 228–242. Springer, Heidelberg (2010)

30. Wiederhold, G.: Mediators in the architecture of future information systems. IEEE Comput. **25**(3), 38–49 (1992)

31. Yuan, P., Liu, P., Wu, B., Jin, H., Zhang, W., Liu, L.: Triplebit: a fast and compact system for large scale RDF data. PVLDB **6**(7), 517–528 (2013)

32. Zadorozhny, V., Raschid, L., Vidal, M.-E., Urhan, T., Bright, L.: Efficient evaluation of queries in a mediator for websources. In: Proceedings of the SIGMOD Conference, pp. 85–96 (2002)

YAM: A Step Forward for Generating a Dedicated Schema Matcher

Fabien Duchateau[1](✉) and Zohra Bellahsene[2]

[1] Université Lyon 1, LIRIS UMR 5205, Lyon, France
fabien.duchateau@univ-lyon1.fr
[2] Université Montpellier, LIRMM, Montpellier, France
bella@lirmm.fr

Abstract. Discovering correspondences between schema elements is a crucial task for data integration. Most schema matching tools are semi-automatic, e.g., an expert must tune certain parameters (thresholds, weights, etc.). They mainly use aggregation methods to combine similarity measures. The tuning of a matcher, especially for its aggregation function, has a strong impact on the matching quality of the resulting correspondences, and makes it difficult to integrate a new similarity measure or to match specific domain schemas. In this paper, we present YAM (Yet Another Matcher), a matcher factory which enables the generation of a dedicated schema matcher for a given schema matching scenario. For this purpose we have formulated the schema matching task as a classification problem. Based on this machine learning framework, YAM automatically selects and tunes the best method to combine similarity measures (e.g., a decision tree, an aggregation function). In addition, we describe how user inputs, such as a preference between recall or precision, can be closely integrated during the generation of the dedicated matcher. Many experiments run against matchers generated by YAM and traditional matching tools confirm the benefits of a matcher factory and the significant impact of user preferences.

Keywords: Schema matching · Data integration · Matcher factory · Schema matcher · Machine learning · Classification

1 Introduction

There are a plethora of schema matching tools designed to help automate what can be a painstaking task if done manually [3]. The diversity of tools hints at the inherent complexity of this problem. The proliferation of schema matchers and the proliferation of new (often domain-specific) similarity measures used within these matchers have left data integration practitioners with the very perplexing task of trying to decide which matcher to use for the schemas and tasks they need to solve. Traditionally, the matcher, which combines various similarity measures, is based on an aggregation function. Most matching tools are semi-automatic,

© Springer-Verlag Berlin Heidelberg 2016
A. Hameurlain et al. (Eds.): TLDKS XXV, LNCS 9620, pp. 150–185, 2016.
DOI: 10.1007/978-3-662-49534-6_5

meaning that to perform well, an expert must tune some (matcher-specific) para-meters (thresholds, weights, etc.) Often this tuning can be a difficult task as the meaning of these parameters and their effect on matching quality can only be seen through trial-and-error [39]. Lee et al. have shown how important (and dif-ficult) tuning is, and that without tuning most matchers perform poorly, thus leading to a low quality of the data integration process [27]. To overcome this, they proposed *eTuner*, a supervised learning approach for tuning these matching knobs. However, eTuner has to be plugged into a matching tool, which requires programming skills. A user must also still commit to one single matcher (the matcher provided in the matching tool). Several research papers [10,12,18,31] led to the conclusion that matchers based on machine learning provide acceptable results w.r.t. existing tools. The main idea consists of training various similarity measures with a sample of schemas and correspondences, and applying them to match another set of schemas. Our intuition is that machine learning can be used at the matcher level.

Another motivation deals with the pre-match interaction with the users [38]. They usually have some preferences or minor knowledge of the schemas to be matched, which are rarely used by the schema matchers [43]. For instance, a quick examination of the schemas may have revealed a few correct correspon-dences, or the user may have an external resource (dictionary, ontology) or a dedicated similarity measure which could be exploited for a specific domain. Schema matchers (often implicitly) are designed with one or a few matching tasks in mind. A matcher designed for automated web service composition may use very stringent criteria in determining a match, i.e., it may only produce a correspondence if it is close to 100 % confident of the correspondence's accuracy. In other words, such a matcher uses precision as its performance measure. In contrast, a matcher designed for federating large legacy schema may produce all correspondences that look likely, even if they are not certain. Such a matcher may favor recall over precision, because the human effort in "rejecting" a bad corre-spondence is much less than the effort needed to search through large schemas and find a missing correspondence. This difference can make a tremendous dif-ference in the usefulness of a matcher for a given task. Integrating these user preferences or knowledge prior to the matching is a challenge for improving the quality results of a schema matcher.

In this context, we present **YAM**, which is actually not *Yet Another Matcher*[1]. Rather YAM is the first schema matcher generator designed to pro-duce a tailor-made matcher, based on the automatic tuning of the matcher and the optional integration of user requirements. While schema matching tools pro-duce correspondences between schemas, YAM is a matcher factory because it produces a dedicated schema matcher (that can be used later for discovering correspondences). This means that theoretically, YAM could generate schema matchers which are very similar to the tools COMA++ [2] or MatchPlan-ner [18]. Schema matching tools only have one predefined method for combining similarity measures (e.g., a weighted average), while a schema matcher generated

---

[1] The name of the tool refers to a discussion during a panel session at XSym 2007.

by YAM includes a method selected among many available (e.g., a decision tree, a Bayesian network, a weighted average). To fulfill this goal, YAM considers the schema matcher as a machine learning classifier: given certain features (i.e., the similarity values computed with various similarity measures), the schema matcher has to predict the relevance of a pair of schema elements (i.e., whether this pair is a correspondence or not). In this framework, any type of classifier can be trained for schema matching, hence the numerous methods available for combining similarity measures. YAM does not only select the best method for combining similarity measures but it also automatically tunes the parameters inherent to this method (e.g., weights, thresholds). The automatic tuning capability has been confirmed as one of the challenges proposed in [43]. In addition, YAM integrates user preferences or knowledge (about already matched scenarios and training data), if available, during the generation of the schema matchers. For instance, our approach benefits from expert correspondences provided as input, because they are used for generating the schema matcher. In YAM, we also allow a user to specify her/his preference for precision or recall, and we produce a dedicated matcher that best meets the users needs. YAM is the first tool that allows the tuning of this very important performance trade-off.

The main contributions in this paper are:

- Our approach is the first to refer to **schema matching as a classification task**. Although other approaches may use classifiers as a similarity measure, we propose to consider a schema matcher which combines various similarity measures as a classifier.
- In addition, our work is the first **matcher factory** for schema matching. Contrary to traditional matching approaches, our factory of matchers generates different matchers and selects the dedicated matcher for a given scenario, i.e., the matcher which includes the most relevant similarity measures, combined with the most appropriate method, and best tuned.
- Another contribution deals with the **close integration of user preferences** (e.g., preference between precision or recall, expert correspondences).
- A **tool named YAM** has been implemented. Its main features include self-tuning (the method for combining similarity measures in a matcher is tailored to the schemas to be matched) and extensibility (new similarity measures or classifiers can be added with no need for manual tuning).
- Experiments over well-known datasets were performed to **demonstrate the significant impact of YAM** at different levels: the need for a matcher factory, the benefit of user preferences on the matching quality, and the evaluation with other schema matching tools in terms of matching quality and performance time.

Outline. The rest of the paper is organized as follows. Section 2 contains the necessary background and definitions of the notions and concepts that are used in this paper. Section 3 gives an overview of our approach while Sect. 4 provides the details of the learning process. The results of experiments showing the effectiveness of our approach are presented in Sect. 5. Related work is described in Sect. 6. Finally, we conclude in Sect. 7.

2   Preliminaries

Schema matching is traditionally applied to matching pairs of edge-labeled trees (a simple abstraction that can be used for XML[2] schemas, web interfaces, JSON[3] data types, or other semi-structured or structured data models). The schema matching task can be divided into three steps. The first one is named **pre-match** and is optional. Either the tool or the user can intervene, for instance to provide resources (dictionaries, expert correspondences, etc.) or to set up parameters (tuning of weights, thresholds, etc.). Secondly, the **matching process** occurs, during which correspondences are discovered. The final step, the **post-match process**, mainly consists of validation of the discovered correspondences by the user.

Definition 1 (Schema): A schema is a labeled unordered tree $S = (E_S, D_S, r_S, label)$ where E_S is a set of elements; r_S is the root element; $D_S \subseteq E_S \times E_S$ is a set of edges; and $label\ E_S \rightarrow \Lambda$ where Λ is a countable set of labels.

Definition 2 (Schema matching scenario): A schema matching scenario is a set of schemas (typically from the same domain, e.g., *genetics* or *business*) that need to be matched. A scenario may reflect one or more properties (e.g., domain specific, large scale schemas). An example of schema matching scenario is composed of two *hotel booking* web forms, such as those depicted by Fig. 1(a) and (b). Optionally, a schema matching scenario can include user preferences (preference for precision or recall, expert correspondences, scenarios from the same domain, number of training data and choice of the matching strategy). These options are detailed in Sect. 4.3. In the next definitions, we focus on a scenario with two schemas S_1 and S_2 for clarity, but the definitions are valid for a larger set of schemas.

Definition 3 (Dataset): A dataset is a set of schema matching scenarios. For instance, the dataset used in the experiments of this paper is composed of 200 scenarios from various domains.

Definition 4 (Pair): A pair of schema elements is defined as a tuple $<e_1, e_2>$ where $e_1 \in E_1$ and $e_2 \in E_2$ are schema elements. For instance, a pair from the two *hotel booking* schemas is $<city, hotel\ name>$.

Definition 5 (Similarity Measure): A similarity measure is a function which computes a similarity value between a pair of schema elements $<e_1, e_2>$. The similarity value is noted $sim(e_1, e_2)$ and it indicates the likeness between both elements. It is defined by:

$$sim(e_1, e_2) \rightarrow [0, 1]$$

---

[2] Extensible Markup Language (XML) (November 2015).
[3] JavaScript Object Notation (JSON) (November 2015).

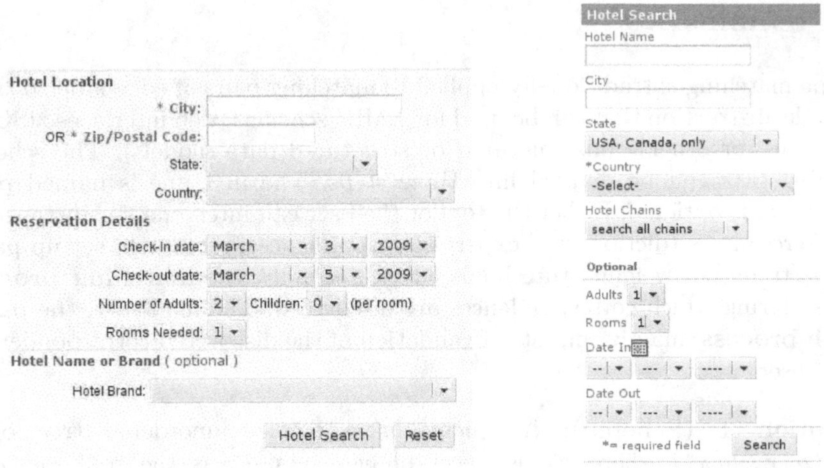

(a) Web form *hotels-valued*

(b) Web form *where-to-stay*

Fig. 1. Two web forms about hotel booking

where a *zero* value means total dissimilarity and a value equal to *one* stands for total similarity. Note that measures computed in $\Re$ can usually be converted in the range $[0, 1]$. In the last decades, many similarity measures have been defined [20,25,29] and are available in libraries such as Second String[4].

Definition 6 (Correspondence): A correspondence is a pair of schema elements which are semantically similar. It is defined as a tuple $<e_1, e_2, k>$, where k is a confidence value (usually the average of all similarity values computed for the pair $<e_1, e_2>$). A set of correspondences can be provided by an expert (ground truth) or it may be produced by schema matchers. Figure 2 depicts two sets of correspondences. The set on the left side (Fig. 2(a)) is the expert set, which includes expected correspondences. The set on the right side (Fig. 2(b)) has been discovered by a schema matcher (YAM). Note that an expert correspondence traditionally has a similarity value equal to 1. As an example, $<searchform, search, 1>$ is an expert correspondence.

Definition 7 (Schema Matcher): A schema matcher is an algorithm or a function which combines similarity measures (e.g., the average of similarity values for a pair of schema elements). In addition, a matcher includes a decision step to select which pair(s) become correspondences (e.g., the decision may be a threshold or a top-K). Given a pair $<e_1, e_2>$ and its similarities values computed for k similarity measures, we represent the combination *comb* and the decision

---

[4] Second String (November 2015).

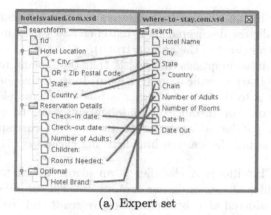

(a) Expert set

(b) YAM's set (using a dedicated matcher)

Fig. 2. Two sets of correspondances for the hotel booking example

dec of a matcher as follows:

$$dec(comb(sim_1(e_1, e_2), \dots, sim_k(e_1, e_2))) \rightarrow <e_1, e_2, k> \text{ or } \emptyset$$

A few matchers also use information about other pairs (e.g., $<e_1, e_3>$) to decide whether the pair $<e_1, e_2>$ is a correspondence or not. Thus, a more generic definition of a schema matcher M focuses on the fact that it produces a set of correspondences between two schemas S_1 and S_2:

$$M(S_1, S_2) \rightarrow \{<e_1, e_2, k>\}$$

Definition 8 (Schema matching quality): The schema matching quality evaluates the effectiveness of a schema matcher for a given schema matching scenario, by comparing the correspondences discovered by this matcher for the scenario to a ground truth. Figure 2(a) provides the ground truth between the two web forms. Different metrics have been designed to measure the effectiveness of a matcher. For instance, precision measures the rate between the number of correct correspondences discovered by a matcher and the number of expected correspondences (provided in the ground truth).

One of the main assumptions in YAM is that the schema matching process can be seen as a binary classification algorithm. Indeed, the goal of the schema matching process is to determine whether a pair of schema elements is a correspondence or not. For instance, a matcher will have to classify the pair $<$*city, hotel name*$>$ either as a correct or an incorrect correspondence. Thus, YAM views schema matchers as machine learning classifiers [34].

Definition 9 (Classifier): A classifier is an algorithm that aims at sorting input data into different classes. In our context, two classes indicate the validity of a pair to be considered as a correspondence: *relevant* and *irrelevant*. Different types of classifiers are available such as decision trees or support vector machines [23,34]. Two processes are usually associated to a classifier: *training* (or learning) consists of building a classifier from a given type by exploiting training data while *using* stands for the application of the trained classifier against another dataset. To train a classifier[5], training data is described with features and a class. In our context, the set of training data τ is represented with similarity measures and associated values and its validity v:

$$\tau = \{((sim_1, sim_1(e_1, e_2)), \ldots, (sim_k, sim_k(e_1, e_2)), v)\}$$

Given a set of training data τ, the training for a type of classifier ω produces a classifier c as follows:

$$\omega(\tau) \to c$$

When training a classifier, the main objective is to promote a given evaluation metric, and the chosen metric for YAM is the misclassification rate. At the end of the learning, the generated classifier efficiently combines (a subset of) similarity measures. In our context, using a classifier is equivalent to a schema matcher, i.e., it produces a set of schema correspondences between two input schemas:

$$c(S_1, S_2) \to \{<e_1, e_2, k>\}$$

Definition 10 (Matcher factory): A factory of matchers such as YAM produces different schema matchers based on the same inputs. Each of those matchers has its own specificity, mainly for combining similarity measures, tuning internal parameters, taking a decision. In our context, these specificities mainly relate to the types of classifiers. Given two schemas S_1 and S_2, a set of training data τ, a set of type of classifiers Ω, and optional user preferences Φ, a matcher factory generates a set of generated matchers $\mathcal{C}$, i.e., one matcher for each type of classifier.

$$M(S_1, S_2, \tau, \Omega, \Phi) \to \mathcal{C} \qquad \text{where } \mathcal{C} = \{c_1, \ldots, c_n\} \text{ and } |\Omega| = n$$

---

[5] We focus on supervised classification, i.e., all training data are labelled with a class.

Definition 11 (Dedicated matcher): The motivation for generating many classifiers within a factory comes from the fact that a given schema matcher, even craftily tuned, may not reach the quality level of another matcher for a given scenario [27]. Yet, no schema matcher performs well in all possible scenarios. In addition to generating many matchers, a factory of schema matchers is also in charge of selecting the dedicated schema matcher, i.e., the "best matcher among all those generated". The notion of "best matcher" depends on a strategy which encompasses three criteria: an evaluation metric, a validation dataset and a pool of matchers. Strategies are described in more detail in Sect. 4.4. Broadly speaking, given an evaluation metric μ, a set of matchers $\mathcal{C}$ and a validation dataset $\mathcal{X}$, the dedicated matcher $\Gamma \in \mathcal{C}$ is the classifier which obtains the highest score for the evaluation metric against the validation dataset:

$$\forall c_i \in \mathcal{C}, \mu(\Gamma, \mathcal{X}) \geq \mu(c_i, \mathcal{X})$$

This dedicated matcher Γ is then used for matching S_1 and S_2.

3 Overview of Our Approach

YAM is a self-tuning and extensible matcher factory tool, which generates a dedicated schema matcher according to a scenario and optional user requirements. Broadly speaking, YAM includes a repository of training data (scenarios with expert correspondences) and a set of types of classifier. It generates tuned schema matchers for various types of classifier, and then select the "best" one - according to a strategy - as the dedicated matcher. This dedicated matcher can be used for matching the input scenario. The **self-tuning** feature stands for the ability to produce a matcher with appropriate characteristics for a given scenario, mainly the method for combining similarity measures (aggregation functions, Bayes network, decision trees, etc.). The **extensible** feature enables users of a matching tool to add new similarity measures. Traditional matching tools which offer this extensibility are often restricted by the manual update of the configuration for both the similarity measures and the method which combines them (e.g., adjusting thresholds, re-weighting values). However, YAM automatically tunes these parameters and is therefore easily extensible. Finally, the **integration of user requirements** allows YAM to convert user time spent to specify these requirements into better quality results, mainly by generating matchers specifically tuned for the scenario. YAM provides these three capabilities because it is based on machine learning techniques, as described in the next part on the architecture. The last part of the section illustrates a running example with YAM.

3.1 Architecture of the YAM System

To the best of our knowledge, YAM is the first factory of schema matchers and it aims at generating a dedicated schema matcher (i.e., a craftily tuned schema matcher) for a given schema matching scenario. For this purpose, YAM uses machine learning techniques during the pre-match phase.

Figure 3 depicts the architecture of YAM. The circles represent inputs or outputs and the rectangles stand for processes. Note that a dotted circle means that such an input is optional. YAM requires only one input, the set of schemas to be matched. However, the user can also provide additional inputs, i.e., preferences and/or expert correspondences (from a domain of interest, or for the input schemas to be matched). The preferences consist of a precision and recall trade-off and a strategy to select the dedicated matcher. In YAM, a repository stores a set of classifiers (currently 20 from the Weka library [23]), a set of similarity measures (mainly from the Second String project [41]), a set of training data (200 schema matching scenarios from various domains with their expert correspondences). The **schema matcher generator** is in charge of generating one tuned matcher for each classifier in the repository according to the user inputs (see Sects. 4.2 and 4.3). Then, the **schema matcher selector** applies a strategy to choose the dedicated matcher among all the tuned matchers (see Sect. 4.4). Finally, this dedicated schema matcher can be used for matching, and specifically the input schemas for which it was tailored. This matching process produces a list of correspondences discovered between the input schemas. Note that the matching process is specific to the type of classifier that will be used and it is not detailed in this paper. For instance, MatchPlanner performs the matching with a decision tree [18] while SMB uses the Boosting meta-classifier [31]. Next, we explain how YAM works with a simple example based on the *hotel booking* web forms.

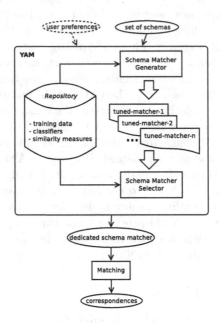

Fig. 3. Architecture of YAM

3.2   Running an Example

A user needs to match two schemas for *hotel booking*. The user is not an expert in data integration, and does not have any idea about the appropriate similarity measures or the configuration of the parameters for matching these schemas. By using YAM, the user simply provides the two schemas as input and runs the matching. Since no preferences have been provided, YAM has a default behaviour and it uses random training data from its repository for learning the dedicated matcher. First, YAM generates and tunes one schema matcher for each type of classifier. Among these generated schema matchers, the one with the best results on the training data is elected as the dedicated matcher. Let us imagine that the dedicated matcher is based on a *Bayes Net* type of classifier. YAM can finally use this dedicated matcher to match the schemas of the *hotel booking* scenario (Fig. 1). The matching phase depends on the type of classifier of the dedicated matcher (*Bayes Net*). In this example, the dedicated matcher computes the probability of a pair of schema elements being a correspondence for each similarity measure. A global probability is finally derived to indicate whether the pair is a correspondence or not. The set of correspondences resulting from this matching is shown in Fig. 2(b). In comparison with the ground truth (Fig. 2(a)), we notice that 8 out of the 9 correct correspondences have been discovered. However, two irrelevant correspondences have also been found, namely (*Hotel Location*, *Hotel Name*) and (*Children:*, *Chain*).

In this simple motivating example, we have described the main workflow of YAM. The next section describes the learning process in detail by including the integration of user requirements, which reduces the impact of the random training data.

4   Learning a Dedicated Matcher

In this section, we describe YAM's approach to learning a dedicated matcher for a given matching scenario. This section is organized as follows: the first part explains the relation between classification and schema matching. Then, we describe the matcher factory, or how YAM generates a matcher for each classifier. Next, we detail how user preferences are integrated during the learning process. Finally, we focus on the strategies which aim at selecting the dedicated matcher among all generated matchers.

4.1   Matching as a Machine Learning Task

The machine learning classification problem consists in predicting the class of an object from a set of its attributes [34]. Thus, any schema matcher can be viewed as a classifier. Each pair of schema elements is considered as a machine learning object where its attributes are the similarity values computed by a set of selected similarity measures of these elements. Given the similarity values of a pair, a matcher labels this pair as either *relevant* or *irrelevant* (i.e., as a correspondence or not). Of course, a matcher may use any algorithm to compute its result – classification, clustering, an aggregation of similarity measures, or any number of ad hoc methods including techniques such as blocking to improve its efficiency. In YAM, we use an extensible library of types of classifiers, among which are decision trees (e.g., *J48, NBTree*), aggregation functions (e.g., *SimpleLogistic*), lazy classifiers (e.g., *IBk, K**), rule-based classifiers (e.g., *NNge, JRip*), voting systems or Bayes Networks. Three assumptions are required to include a type of classifier in YAM: first, the type of classifier should support supervised learning, i.e., all training data have to be labelled. Secondly, it has to use both numerical and categorical features, i.e., the similarity measures may return either numerical values or semantic values (e.g., *synonym*). The last assumption deals with the discretization ability [15,21], i.e., the type of classifier should be able to split values for continuous features (e.g., similarity measures which return a distance or a value in [0, 1]). The generation of a dedicated matcher can be divided into two steps: (i) training of tuned matchers, which can be impacted by parameters and user preferences, and (ii) selection of the dedicated matcher.

Example: Let us consider the pair *(searchform, search)* from our running example. We computed the similarity values of this pair for each similarity measure in our library. For instance, let us assume we have three similarity measures: *AffineGap* [1], *NeedlemanWunsch* [36] and *JaroWinkler* [45]. Processing them over the pair *(searchform, search)* provides the following similarity values: *AffineGap* = 14.0, *NeedlemanWunsch* = −4.0, *JaroWinkler* = 0.92. From these values, a matcher must predict whether the pair is a relevant correspondence or not.

4.2 Training Tuned Matchers

In this part, we explain how to generate tuned matchers which aim at classifying pairs in a class, either relevant or irrelevant. To reach this goal, classifiers have to be trained. YAM trains each matcher using its repository of training data and potential expert correspondences provided by the user (see Sect. 4.3). The training data in the repository consists of expert correspondences, i.e., pairs of schema elements with their relevance[6]. The training algorithm is specific to each classifier [34]. However, we shall briefly sum up the main intuition: first, an algorithm selects the similarity measures which provide a maximum of correctly classified correspondences (i.e., a minimal misclassification rate). Then, the similarity measures that might solve harder cases are taken into account.

To illustrate the training process, we have chosen a well-known classifier, the decision tree. In our context, a decision tree is a tree whose internal nodes represent the similarity measures, and the edges stand for conditions applied to the result of the similarity measure (e.g., the similarity value computed by a measure must be greater than a value). All leaf nodes in the tree are the classes, either a *relevant correspondence* or an *irrelevant correspondence*. Algorithm 1 describes the learning of a decision tree in our context. The learning takes as

---

[6] The two schemas of a pair may be necessary to compute similarity values, for instance with structural or contextual measures.

input a set of training data $\mathcal{T}$ and a set $\mathcal{S}$ containing k similarity measures. This training data is defined as a set $\mathcal{T} = \{t\}$, and a single training data $t_i \in \mathcal{T}$ is represented as $t_i = \{(sim_{i1}, v_{i1}), \ldots, (sim_{ik}, v_{ik}), (label_i, class_i)\}$. The output is a decision tree. In the initialization function $buildDecisionTree$, an empty decision tree is created and the recursive function $partition$ is called (lines 2 and 3). The goal of this second function is to split the training data into one or more classes, thus creating a new level in the decision tree. To fulfill this goal, the best feature has to be selected for partitioning the training data. Note that similarity measures are continuous features (values in the range $[0, 1]$) and they need to be discretized. This discretization is a well-known problem [15,21] and it generates cut points, e.g., conditions representing a range of values associated to a class. For each similarity measure, the algorithm produces a set of cut points (lines 6 to 11). Each training data is then associated to one class, i.e., the training data satisfies the condition of a given cut point (lines 12 to 19). As a result, $\mathcal{P}_{sim}^{class}$ contains the training data of the class $class$ according to a given feature sim. When the partitioning has been performed for all similarity measures, the algorithm is able to select the best partition according to an evaluation metric (line 24). Various evaluation metrics are available, such as information gain, Gini criterion or gain ratio [46], and we rely on misclassification rate in our context. Finally, if the partition produced by the best similarity measure only contains one class[7], then there is no more partitioning of the data and that single class is added as a child node in the tree (line 26). If several classes are present in the partition of the best similarity measure, then the algorithm adds each of these classes as child nodes in the tree, and the function $partition$ is recursively called for each class and its training data (lines 28 to 31). Note that Algorithm 1 aims at facilitating the understanding of the learning process, but the building of a classifier is usually improved with heuristics such as pruning [34].

Example: Let us study an example for generating a decision tree with this algorithm. The training data is composed of nine pairs of elements, among which three are relevant, namely $<searchform, search>$, $<city, city>$ and $<brand, chain>$. Two similarity measures, $Trigrams$ and $Context$, serve as attributes. Figure 4 depicts the generation of the decision tree at the first iteration. A matrix represents the classification performed by a measure, and a pair is either classified as relevant (R) or irrelevant (I). The **white** background colour (respectively **grey**) indicates that a pair is correctly (respectively incorrectly) classified. For instance, the pair $<searchform, city>$ has been correctly classified as an irrelevant correspondence by the $Trigrams$ measure (Fig. 4(a)) while it has been incorrectly classified as a relevant correspondence by the $Context$ measure (Fig. 4(b)). Note that each classifier is in charge of adjusting the thresholds of each similarity measure to obtain the best classification. Given these matrices, the misclassification rate ϵ is computed for each measure (by counting the number of grey boxes). In this case, $Trigrams$ has achieved the lowest misclassification rate ($\frac{2}{9}$) and it is therefore selected to be added to the decision tree,

---

[7] Other stop conditions may be used, for instance "all training data have been correctly classified".

Algorithm 1. Algorithm for building a decision tree

Input: set of training data $\mathcal{T}$, set of similarity measures $\mathcal{S}$
Output: a decision tree $\mathcal{D}$
1: **function** buildDecisionTree($\mathcal{T}$, $\mathcal{S}$)
2:   $\mathcal{D} \leftarrow \emptyset$
3:   partition($\mathcal{T}$, $\mathcal{S}$, $\mathcal{D}$)
4: **end function**
5:
6: **function** partition($\mathcal{T}$, $\mathcal{S}$, $parent$)
7:   $\mathcal{P} \leftarrow \emptyset$
8:   **for all** sim $\in \mathcal{S}$ **do**
9:     $\mathcal{P}_{sim} \leftarrow \emptyset$
10:    $\mathcal{T}_{sim} \leftarrow \mathcal{T}$
11:    $\mathcal{CP} \leftarrow$ discretize($\mathcal{T}_{sim}$, $\mathcal{S}$)
12:    **for all** ($condition$, $class$) $\in \mathcal{CP}$ **do**
13:      $\mathcal{P}_{sim}^{class} \leftarrow \emptyset$
14:      **for all** $t \in \mathcal{T}_{sim}$ **do**
15:        **if** $t \vdash condition$ **then**
16:          $\mathcal{P}_{sim}^{class} \leftarrow \mathcal{P}_{sim}^{class} \cup \{t\}$
17:          $\mathcal{T}_{sim} \leftarrow \mathcal{T}_{sim} - \{t\}$
18:        **end if**
19:      **end for**
20:      $\mathcal{P}_{sim} \leftarrow \mathcal{P}_{sim} \cup \{\mathcal{P}_{sim}^{class}\}$
21:    **end for**
22:    $\mathcal{P} \leftarrow \mathcal{P} \cup \{\mathcal{P}_{sim}\}$
23: **end for**
24: $best\_sim =$ findBestClassification($\mathcal{P}$)
25: **if** $-\mathcal{P}_{best\_sim}- = 1$ **then**
26:    addChild($parent$, $class$)
27: **else**
28:    **for all** $class \in \mathcal{P}_{best\_sim}$ **do**
29:      addChild($parent$, $class$)
30:      partition($\mathcal{P}_{best\_sim}$, $\mathcal{S}$, $class$)
31:    **end for**
32: **end if**
33: **end function**

---

as shown in Fig. 4(c). The variables X_1 and X_2 stand for the threshold values which enable the achievement of this best classification. At the end of the first iteration, two pairs were not correctly classified by the *Trigrams* measure. Thus, a new iteration begins in order to classify these two pairs with all similarity measures. The matrices for the second iteration are shown in Fig. 5(a) and (b). Since the training data are now composed of two pairs at the second iteration, the classifier proposes different threshold and parameter values for each similarity measure. This time, the *Context* measure has correctly minimized the misclassification rate and it is promoted in the decision tree, as shown in Fig. 5(c). Since all the training data have been classified, the algorithm stops.

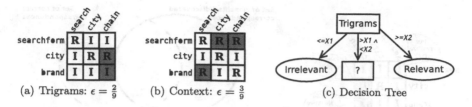

Fig. 4. Training of a decision tree at first iteration

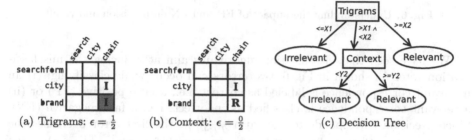

Fig. 5. Training of a decision tree at second iteration

During the training phase, all the thresholds, weights, and other parameters of the matcher (i.e., classifier) are automatically configured, thus providing tuned matchers. Next, we study how user preferences are integrated during this training phase.

4.3   Integrating User Preferences

We have identified five options that the user may configure: (i) **preference for precision or recall**, (ii) **expert correspondences**, (iii) **scenarios from the same domain** (iv) **number of training data** and (v) **strategy** to select a dedicated matcher. These options can be combined to improve the matching quality. We should keep in mind that the user has no requirement to provide options, and specifically the training data and the strategy which are automatically selected by YAM when necessary.

Preference for Precision or Recall. The ability to promote either precision or recall is the first attempt to leverage the results of the matching quality. Many applications need such tuning. For instance, matching tools may require training data (usually expert correspondences as in Glue [13]), and YAM could automatically discover a few correct correspondences by generating a high-precision matcher. On the other hand, a typical scenario in which a high recall is necessary is a matching process followed by a manual verification. Since the validation of a discovered correspondence is cheaper in terms of time and resources than the search of a missing correspondence, the discovery of a maximum number of correct correspondences is crucial and implies a high-recall matcher [17].

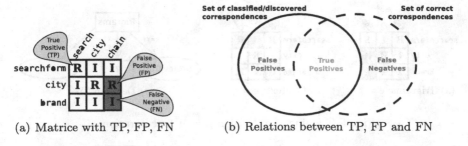

(a) Matrice with TP, FP, FN   (b) Relations between TP, FP and FN

Fig. 6. Understanding the impact of FP and FN on precision and recall

We have seen that classification algorithms aim at reducing the misclassification rate. As shown in Fig. 6, two errors can occur while classifying: (i) an irrelevant correspondence is labeled as correct, i.e., a false positive (FP) or (ii) a relevant correspondence is classified as incorrect, i.e., a false negative (FN). Since precision corresponds to the ratio $\frac{TP}{TP+FP}$, the first error decreases the precision value. Conversely, recall is computed with formula $\frac{TP}{TP+FN}$, thus the second error has a negative impact on recall.

To produce tuned matchers which promote either precision or recall, we propose to set a penalty for false positives or false negatives during the learning. This means that false positives (or false negatives) have a stronger impact when computing the misclassification rate. To increase precision (respectively recall) on a given training dataset, we assign a greater penalty to false positives (respectively false negatives). Note that promoting recall (respectively precision) mainly decreases precision (respectively recall).

Example: Back to the first iteration of our example, but let us imagine that we want to promote recall to avoid the misclassification of relevant correspondences (see Fig. 7). Therefore, a penalty - equal to 4 in this example - is set for false negatives. Due to the false negative <*brand, chain*>, the misclassification rate of the measure *Trigrams* drops to $\frac{5}{9}$ and the *Context* measure is selected to be added in the decision tree with its two threshold values Y_1 and Y_2. The three false positives of the *Context* measure should be reclassified in the next iteration. In that way, YAM is able to produce matchers which favour the preference of the user in terms of matching quality.

Expert Correspondences. The training data mainly consist of correspondences from the repository. However, the user may decide to provide expert correspondences between the schemas to be matched[8]. The benefit for providing these correspondences is threefold. First, expert correspondences enable a better tuning for some similarity measures. For instance, structural or contextual measures analyse the neighbouring elements in the schema for each element

---

[8] If the user has not provided a sufficient number of correspondences, YAM will extract others from the repository.

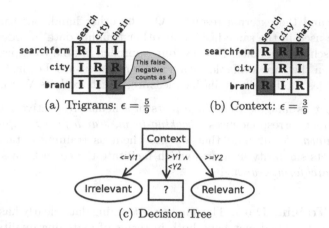

(c) Decision Tree

Fig. 7. Training of a decision tree at first iteration while promoting recall

of the correspondence in order to evaluate their similarity. Such measures are given more weight in this case. Another benefit is specific to the constraint-based measures which may discard candidate pairs by relying on the relevance of expert correspondences. Finally, the logic for designing a schema is usually kept throughout the whole process. An expert correspondence between the two schemas reflects the fact that the similarity measure(s) which confirm this correspondence may have captured (a part of) this logic. These measures can be useful for assessing the relevance of other pairs between the same schemas. Section 5.3 includes experiments showing the impact of these expert correspondences on the matching quality.

Example: Let us illustrate this impact with our running example. The consequence of providing the expert correspondence <*searchform, search*> can be explained as follows. The *Context* measure (Figs. 4(b) or 7(b)) analyses the similarity of other pairs (e.g., <*searchform, city*>). Since an expert correspondence is established between <*searchform, search*>, the context of the *searchform* element is modified, and it can imply that the pair <*searchform, city*> is now classified as irrelevant. This means that the *Context* measure can have a lower misclassification rate and thus be promoted as the measure to be added to the decision tree at the first iteration.

Scenarios from the Same Domain. Similarly to expert correspondences, providing scenarios from the same domain as the schemas to be matched (e.g., the *hotel booking* domain) produces better tuned matchers. The main reason is the vocabulary which is shared across the domain, thus promoting the relevant similarity measures to detect correspondences . The exploitation of external resources (e.g., domain ontology, thesaurus) for schema matching has already been studied [16,47]. However, these external resources differ from our same-domain scenarios: when using external resources, a matching has to be performed between the schemas and the external resource. On the other hand, our same-domain scenarios are used in the same fashion as other training data in order to generate a tuned schema matcher. Besides, our approach is able to exploit external resources when they are available with a similarity measure. For instance, YAM already includes the Resnik similarity measure, which exploits Wordnet [40].

Example: Let us imagine that three pairs of schemas respectively include the following correct correspondences <*booking-form, search-form*>, <*form, form*> and <*searchform, form*> and that we use them as training data. Then, the classifier selects similarity measures which facilitate the correct classification of the pair <*searchform, search*>.

Number of Training Data. The number of training data clearly has an impact on generating the tuned matchers, both in terms of matching quality and time performance. The common rule is that a classifier performs better when trained with more data. Yet, there are threshold values above which a classifier can provide stable results. To determine these threshold values for generating robust tuned matchers, we have conducted extensive experiments described in Sect. 5.3. Thus, the number of training data is automatically set by YAM, but the user can tune this value.

Strategy. This last preference can be provided by the user to select the dedicated matcher. The available strategies are detailed in the next part.

4.4   Selecting a Dedicated Matcher

The final step of the learning process deals with the selection of the dedicated matcher among all tuned matchers which have been generated by YAM for a given schema matching scenario. This selection depends on the adopted strategy, which can be user-defined or automatically configured by YAM. The strategy is defined as a combination of one value for each of the three following criteria:

- **Choice of the quality measure**: *precision, recall* or *F-measure*
- **Choice of the validation dataset**: the *repository*, the *set of expert correspondences* or the *similar scenarios* (when provided)
- **Pool of matchers**: *generate* (select the dedicated matcher among the tuned matchers generated in Sect. 4.2), *reuse* (select the dedicated matcher among those stored in the repository), *both* (select the dedicated matcher among the repository and the tuned matchers).

If the user has not manually set up a strategy, YAM applies the following algorithm to configure the strategy:

- If a preference for *precision* or *recall* is set, this preference is the quality measure. Otherwise, the default measure is the *F-measure*.

- If a *set of expert correspondences* or a *set of similar schemas* is provided, this set becomes the validation dataset. Otherwise, YAM uses (a subset of) the scenarios from the *repository*.
- The *generated* matchers form the pool of matchers. The *reuse* and *both* options need to be provided by the user and they mainly aim at speeding up the execution process by avoiding the learning of new tuned matchers.

Once the criteria of the strategy have been fixed, each matcher (from the pool of matchers) computes their accuracy (depending on the selected measure) by performing a cross-validation process against the validation dataset. The matcher which obtains the best value is elected as the dedicated matcher. Note that the dedicated matcher is finally stored in the repository, thus allowing it to be reused for further experiments.

Example: To illustrate the impact of the strategy, let us comment Fig. 8. YAM has generated two tuned matchers (*J48* and *Naive Bayes*) and one matcher from a previous generation is stored in the repository (*Binary SMO*). Each matcher has its own technique for combining the different similarity measures (reflected by thresholds, weights or intern metrics such as standard deviation). To simplify the example, the training data consist only of examples stored in the repository. The boxes below each matcher indicate the results of cross-validation achieved by the matcher over the training data. If the user has not provided any strategy or preference, YAM automatically selects the matcher among those *generated* with the *best F-measure* value (default strategy equal to <*F-measure, repository, generated*>). In this case, the dedicated matcher will be *Naive Bayes* (83 % F-measure) to the detriment of *J48*. On the contrary, if the user has set a preference for recall, the strategy is defined as <*recall, repository, generated*>. Since *J48* obtains a 89 % recall, it is selected as the dedicated matcher. Finally, if the strategy is <*precision, repository, both*>, this means that the user needs a matcher with the best precision value among all matchers both generated and stored in the repository. In this context, *Binary SMO* achieves the best precision value (100 %) and it will be elected as the dedicated matcher.

5   Experiments

This section begins with a description of the protocol. Next, we firstly demonstrate the need for a matcher factory (self-tuning feature). Then we study the integration of user preferences (described in Sect. 4.3.), and their impact on the matching quality. Finally, we compare our results with two matching tools that have excellent matching quality, COMA++ [2] and Similarity Flooding (SF) [32]. These tools are described in more detail in Sect. 6.

5.1   Experimental Protocol

Experiments were run on a 3.6 GHz computer with 4 Go RAM under Ubuntu 11.10.

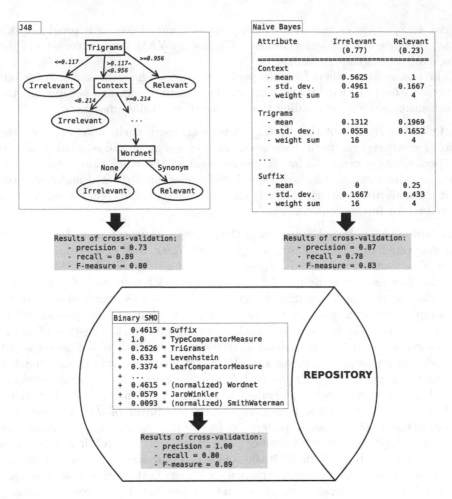

Fig. 8. Results of cross-validation for different matchers

Configurations of the Tools. The default configuration for SF was used in the experiments. We tested the three pre-configured strategies of COMA++ (*AllContext, FilteredContext* and *Fragment-based* in the version *2005b*) and we kept the best score among the three.

The current version of YAM is implemented in Java 1.6. Our tool includes 20 classifiers from the Weka library [23] and 30 similarity measures, including all terminological measures [8] from the Second String project[9], a contextual measure named Approxivect [19], the Resnik semantic similarity measure [40] and a simple structural measure that compares the constraints and data types, as described in Similarity Flooding [32]. YAM's repository contains a large set of 200 schema matching scenarios from various domains.

---

[9] Second String (November 2015).

Dataset. The dataset used in these experiments is composed of more than 200 schema matching scenarios, covering the following domains:

- **University department** describes the organization of university departments [20]. These two small schemas have very heterogeneous labels.
- **Thalia courses.** These 40 schemas have been taken from Thalia collection [24] and they are widely used in literature [14,19]. Each schema has about 20 elements and they describe the courses offered by some worldwide universities. As explained in [44], this dataset could refer to a scenario where users need to generate an exchange schema between various data sources.
- **Travel** includes 5 schemas that have been extracted from airfare web forms [37]. In data sharing systems, partners have to choose a schema or a subset of schema that will be used as a basis for exchanging information. This *travel* dataset clearly reflects this need, since schema matching enables data sharing partners to identify similar concepts that they are willing to share.
- **Currency** and **sms** datasets are popular web services[10]. Matching the schemas extracted from web services is a recent challenge to build new applications such as mashups or to automatically compose web services.
- **Web forms** are a set of 176 schemas, extracted from various websites by the authors of [31]. They are related to different domains, from hotel booking and car renting to dating and betting. For instance, the *finance* domain contains more than ten schemas of small size. Authors of [44] state that schema matching is often a process which evaluates the costs (in terms of resources and money) of a project, thus indicating its feasibility. These scenarios can be a basis for project planning, i.e., to help users decide if integrating their data sources is worth or not.

Table 1 summarizes the features of the schema matching scenarios. The *size* column indicates the average number of schema elements in the scenario. The *structure* column checks how deep the schema elements are nested. We consider a schema to be flat when it includes at most three levels, and a schema is said to be nested with at least four levels. The last column provides information about the number of schemas in the scenario.

For all these scenarios, the expert correspondences are available, either manually or semi-automatically designed. We use these 200 scenarios, and their correct correspondences, both to train YAM and to demonstrate the effectiveness of the three matching tools.

Quality Metrics. To evaluate the matching quality, we use common metrics in the literature, namely precision, recall and F-measure [3,17,20]. Precision calculates the proportion of relevant correspondences extracted among those discovered. Another typical metric is recall which computes the proportion of relevant discovered correspondences between all relevant ones. F-measure is a trade-off between precision and recall.

---

[10] Free Web Services (November 2015).

Table 1. Schema matching scenarios according to their properties

	Average size		Structure		Number of schemas
	Small (<10)	Average (10–100)	Flat (≤3)	Nested (>3)	
Univ. dept	×		×		2
Thalia courses	×	×	×		40
Travel	×		×		5
Currency		×		×	2
Sms		×		×	2
Web forms	×	×	×		176

5.2 Self-tuning Feature

We begin with a study of the self-tuning feature, i.e., the ability to select the most effective schema matcher. More specifically, we justify the need for a schema matcher factory, since our approach can adapt the method for combining similarity measures to the scenario. In other words, if a traditional schema matching tool (e.g., COMA++) performs matching for these 200 scenarios, the same method for combining similarity measures would be used (i.e., an aggregation function for COMA++). With YAM, we demonstrate that from one scenario to another, the optimal method is different (i.e., the dedicated schema matcher generated with YAM is based on different types of classifier).

Let us describe the experiment. We ran YAM against 200 scenarios, and we measured two criteria: the number of times (out of 200) that a type of classifier was selected as the dedicated matcher (Fig. 9(a)) and the average F-measure achieved by a type of classifier over the 200 scenarios (Fig. 9(b)). For instance, the type of classifier *VFI* was selected as a dedicated matcher 57 times (out of 200). This type of classifier *VFI* achieves over the 200 scenarios an average F-measure equal to 59 %. For this evaluation, we included no user preference, so all matchers were trained only with the repository (20 random schema matching scenarios) and the dedicated matcher was selected with the default strategy. This process took roughly 1400 s to produce the dedicated matcher for each given scenario. The plots are limited to the to the 14 best types of classifiers.

The first comment for Fig. 9(a) is the diversity of types of classifier which have been selected. There is not one best schema matcher for matching the 200 scenarios, but more than fourteen. This means that a matcher factory, such as YAM, is necessary to cope with the differences in the schema matching scenarios. Secondly, we note that 2 types of classifier, namely *VFI* (Voting Feature Intervals) and *Bay* (Bayes networks), are selected in half of the 200 scenarios. The matchers based on these types of classifiers can be considered as robust because they provide acceptable results in most scenarios in our repository. This trend is confirmed with the second plot (Fig. 9(b)) on which *VFI* and *Bayes Net* achieve the best average F-measure values over the 200 scenarios. Another comment on these plots deals with the aggregation functions, represented by *SLog* (Simple Logistic) and *MLP*. These functions, which are commonly used by traditional matching tools, are selected as dedicated matchers in only a few scenarios. Thus, they do not provide optimal matching quality results in most schema matching

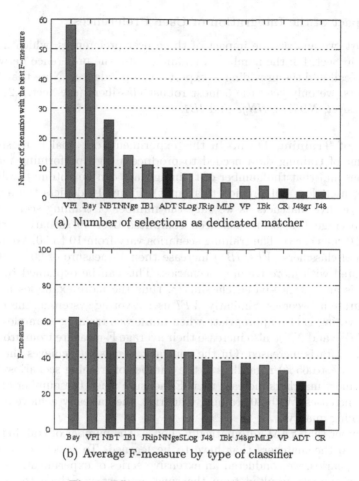

(a) Number of selections as dedicated matcher

(b) Average F-measure by type of classifier

Fig. 9. Effectiveness by type of classifier

scenarios. Finally, a good ranking in terms of F-measure does not guarantee that the type of classifier will be selected many times. For instance, the decision trees *J48* and its alternative *J48graft* obtain an average 40 % F-measure but they are selected as dedicated matchers only a few times. Conversely, the types of classifiers *CR* (Conjunction Rules) and *ADT* (Alternating Decision Tree), which achieve a very low average F-measure on these 200 scenarios (5 % for *CR* and 28 % for *ADT*), were respectively selected 3 and 10 times. This shows that dedicated matchers based on these classifiers are very effective, in terms of matching quality, for solving specific scenarios. Thus, these results support our claim that a matcher factory such as YAM is a promising perspective.

5.3 Impact of the Integration of User Preferences

In this part, we analyse the impact of three user preferences, which have been described in Sect. 4.3: the number of training data, the preference between precision or recall and the providing of expert correspondences. Note that for these experiments, we only keep the 5 most robust classifiers (see Sect. 5.2), namely *VFI*, *BayesNet*, *NBTree*, *NNge* and *IB1*.

Number of Training Data. In this experiment, our goal is to show that the amount of training data needed to produce a high performing matcher is not onerous and that the number of training data can be automatically chosen (when the user does not provide this input). Figure 10 depicts the average F-measure[11] of five matchers as we vary the number of training scenarios. Note that the average F-measure has been computed over 40 scenarios (randomly selected, 20 runs each). The training scenarios vary from 10 to 50. We note that two types of classifiers (*VFI*, *IB1*) increase their F-measure of 20 % when they are generated with more training scenarios. This can be explained by the fact that *IB1* is an instance-based classifier[12], thus the more examples it has, the more accurate it becomes. Similarly, *VFI* uses a voting system on intervals that it builds. Voting is also appropriate when numerous training examples are supplied. *NBTree* and *NNge* also increase their average F-measure from around 10 % as training data is increased. On the contrary, *BayesNet* achieves the same F-measure (60 % to 65 %) regardless of the number of training scenarios. Thus, as expected, most matchers increase their F-measure when the number of training scenarios increases. With 30 training scenarios, they already achieve an acceptable matching quality.

Remember that YAM automatically chooses the number of training scenario according to the matchers that have to be learned. To select this number of training scenarios, we conducted an extensive series of experiments. More than 11, 500 experiments resulted from the runs, and we use them to deduce the number of training scenarios for a given classifier. Table 2 shows the conclusion of our empirical analysis. For instance, when learning a schema matcher based on the *J48* classifier, YAM ideally chooses a number of training scenarios between 20 and 30.

In a machine learning approach, it is crucial to analyse the relationship between performance and the size of training data. Therefore, we evaluate the performance of YAM according to the size of the training data. We have averaged the training and matching times for 2000 runs (10 runs for each of the 200 scenarios) according to different number of training data (from 5 to 50). Table 3 summarizes the results for training 20 classifiers (i.e., 20 tuned matchers), selecting the dedicated matcher, and performing the matching with the dedicated

---

[11] Only the F-measure plot is provided since the plots for precision and recall follow the same trend as the F-measure.

[12] This classifier is named instance-based since the correspondences (included in the training scenarios) are considered as instances during learning. Our approach does not currently use schema instances.

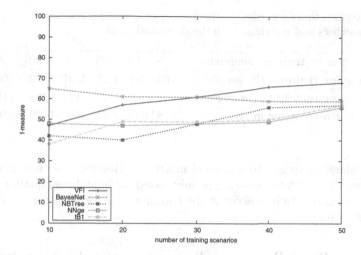

Fig. 10. Average F-measure when varying the number of training scenarios

Table 2. Number of training scenarios for each type of classifier

Number of training scenarios	Classifiers
20 and less	SLog, ADT, CR
20 to 30	J48, J48graft
30 to 50	NNge, JRip, DecTable, BayesNet, VP, FT
50 and more	VFI, IB1, IBk, SMO, NBTree, MLP

matcher. This experiment is independent from the empirical results shown in Table 2, i.e., 20 classifiers were generated with 5 training data, 20 classifiers were generated with 10 training data, etc. This means that the training time (e.g., 165 s for 5 training data) corresponds to the training of 20 classifiers. Obviously some types of classifier are quicker than others to generate a matcher, but our main motivation is the selection of the best tuned matcher among a large panel rather than an efficient generating process. The training step is time-consuming but this is a fair time for learning 20 tuned matchers. The training time seems constant according to the number of training scenarios. The matching time is not significant (between 13 s and up to 128 s). We note that the matching time slightly decreases to 110 s with 50 training scenarios. We believe this is due to the type of classifier which is used: as shown in Table 2, the types of classifiers which are selected with 50 and more training scenarios are mainly instance-based classifiers such as *IB1, VFI* or *IBk*. In our context, the matching with such classifiers seems more efficient. It should be remembered that the training is carried out with 20 classifiers and 30 similarity measures. If required, these numbers can be reduced to improve performance, for instance based on the empirical results from Table 2. Still, an automatic matching performed in around one hour

Table 3. Average times according to the number of training scenarios for both training 20 tuned matchers and matching with the dedicated matcher

Number of training scenarios	5	10	20	30	40	50
Time for training (in seconds)	165	601	2227	3397	5182	6506
Time for matching (in seconds)	13	24	47	124	128	110
Total time (in seconds)	178	625	2274	3521	5310	6616

is an advantage compared to a manual matching. Besides, the current strategy for selecting the dedicated matcher is only based on (matching) quality criteria. But we could also take into account the training time for each type of classifier within the strategy.

Precision vs. Recall Preference. We now present another interesting feature of our tool, the possibility of choosing between promoting recall or precision, by tuning the weight for false positives or false negatives. Schema matching tools usually favour a better precision, but we demonstrate that YAM tuned with a preference for recall effectively allows to obtain a better recall, with no significant impact on F-measure. In other words, the gain in terms of recall is proportionally equivalent to the loss in terms of precision, thus the F-measure is roughly constant. Figure 11(a) and (b) respectively depict the average recall and F-measure of five matchers for 40 scenarios, when tuning the preference between precision and recall. Without any tuning (i.e., weight for false negatives and false positives is equal to 1), this means that we give as much importance to recall as to precision.

For 2 matchers (*NBTree* and *NNge*), the recall increases up to 20 % when we tune in favour of recall. As their F-measures does not vary, it means that this tuning has a negative impact on the precision. However, in terms of post-match effort, promoting recall may be a better choice depending on the integration task for which the matching process is being performed. For example, let us imagine we have two schemas of 100 elements: a precision which decreases by 20 % means a user has to eliminate 20 % of irrelevant discovered correspondences. But a 20 % increase of recall means that (s)he has 20 % fewer correspondences to search through among 10, 000 possible pairs ! Hence, this tuning could have a highly significant effect on the usability of the matcher for certain tasks. Indeed, we highlight the fact that matching tools may be disregarded because the amount of work during pre-match effort (tuning the tool) and the amount of work during post-match effort (manual verification of the discovered correspondences) is sometimes not worthwhile compared to the benefit of the tool, especially if the user cannot leverage the results towards more precision or recall.

For the three other matchers (*BayesNet*, *VFI* and *IB1*), tuning in favour of recall has no significant effect. This does not mean that only a few types of classifiers can promote recall. Without any tuning, only one matcher (*BayesNet*) has an average recall superior to its precision. Indeed, most of the matchers in our

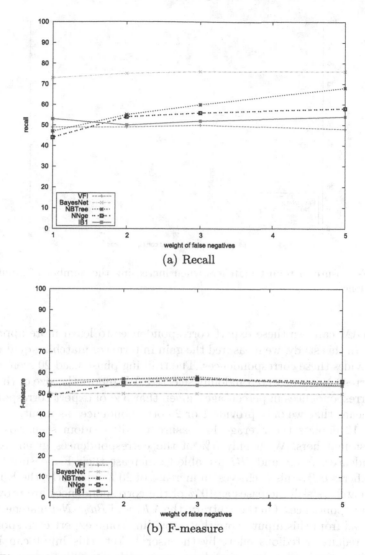

(a) Recall

(b) F-measure

Fig. 11. Matching quality of robust matchers when promoting recall

library promote by default precision. However, when setting a weight for false negatives to 2, then four matchers from the library have a higher recall than precision. And with a weight for false negatives equal to 3, five other matchers reduced the gap between precision and recall to less than 5 %. Thus, this shows how YAM is able to take into account this very important user preference, which directly impacts post-match (manual) effort [17].

Impact of Expert Correspondences. As in Glue [13], the number of expert correspondences is an input - compulsory for Glue, but optional for YAM - to the

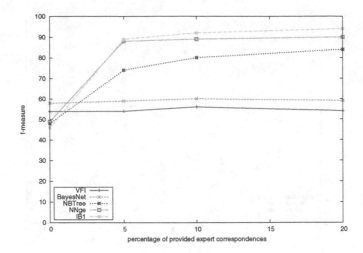

Fig. 12. F-measure of robust matchers when increasing the number of input expert correspondences

system. YAM can use these expert correspondences to learn more appropriate matchers. In this study, we measured the gain in terms of matching quality when a user provides these correspondences. The training phase used 20 scenarios and expert correspondences were randomly selected. We report the size of the sets of expert correspondences in percentages, given that 5 % of expert correspondences usually means that we only provide 1 or 2 correspondences as input.

Figure 12 depicts the average F-measure for 40 random scenarios for the five robust matchers. With only 5 % of the correspondences given as expert correspondences, *NNge* and *IB1* are able to increase their F-measure by 40 %. The classifier *NBTree* also achieves an increase of 20 %. Similarly, the F-measure of these matchers still increases as 10 % of the correspondences are provided as expert correspondences. On the contrary, the *VFI* and *BayesNet* matchers do not benefit at all from this input. Note that providing some expert correspondences does not require a tedious effort by the user[13]. Yet, this input can improve the matching quality of most matchers, even with a small amount of expert correspondences. Besides, YAM closely integrates these expert correspondences in generating a better matcher, while other tools such as Glue mainly use these correspondences as a bootstrap.

5.4   Comparing with Other Matching Tools

In this last experiment, we compare YAM with two matching tools known to provide a good matching quality: COMA++ and Similarity Flooding (SF). COMA++ [2] uses 17 similarity measures to build a matrix between pairs of

---

[13] Some GUIs already exist to facilitate this task by suggesting the most probable correspondences.

elements and aggregate their similarity values. Similarity Flooding [32] builds a graph between input schemas. Then, it discovers some initial correspondences using a string matching measure. These correspondences are refined using a structural propagation mechanism. Both matching tools are described in more detail in Sect. 6. YAM, our factory of schema matchers, uses the default strategy (*<F-measure, repository, generated>*) to produce the dedicated matcher. The number of training data is automatically adjusted according to the classifier which is going to be trained (using Table 2).

Figure 13(a) and (b) depict the F-measure obtained by YAM, COMA++ and Similarity Flooding on 10 schema matching scenarios. YAM obtains the highest F-measure in 7 scenarios, and reaches 80 % F-measure in 4 scenarios. COMA++ achieves the best F-measure for *currency* and *university* scenarios. SF obtains the best F-measure in one scenario (*travel*). Besides, COMA++ is the only tool which does not discover any correspondence for one scenario (*travel*). However, we notice that YAM obtains better results in the web forms scenarios since it was mainly trained with web forms (stored in the repository). With non-web forms scenarios, YAM is still competitive with the other tools.

We have summarized the results of this comparison in Table 4. The numbers in this table represent an average for the 10 scenarios in terms of precision, recall and F-measure. YAM obtains the highest average F-measure (71 %) while COMA++ and SF achieve an average F-measure around 50 %. In addition, in the bottom part of the table we present the matching quality for YAM with user preferences. We note that the promotion of recall is effective (78 % instead of 65 %) but to the detriment of precision. When YAM is trained with scenarios from the same domain, the quality of results slightly improves (F-measure from 71 % to 76 %). The most significant increase in quality is due to the integration of expert correspondences during training, which enables F-measure to reach 89 %.

These experiments show how our matcher factory relies on the diversity of classifiers. Indeed, the dedicated matchers that it has generated for these scenarios are based on various classifiers (*VFI*, *BayesNet*, *J48*, etc.) while COMA++ and SF only rely on respectively an aggregation function and a single graph propagation algorithm. Besides, YAM is able to integrate user preferences to produce more efficient dedicated matchers and to improve the matching quality.

6   Related Work

Much work has been done both in schema matching and ontology alignment. One can refer to the following books and surveys [3,6,20,22,42] for more details about schema and ontology matchers. All related approaches aims at performing matching (or alignment). On the contrary, YAM is a matcher factory, which produces a schema matcher. There is no equivalent approach to our generator of schema matchers. In this section, we have chosen to present an overview of the last decade of research in schema and ontology matching, which has served as a basis to our work. Still, a deeper comparison between traditional matching tools and our factory of matchers is difficult due to the nature of the tools.

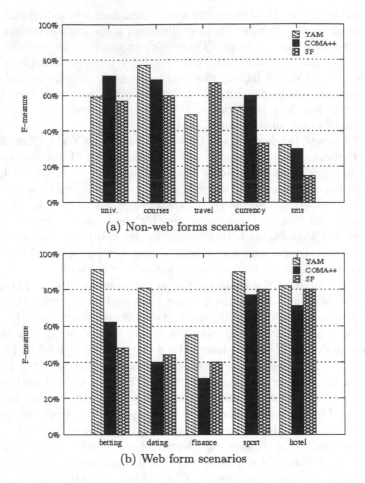

(a) Non-web forms scenarios

(b) Web form scenarios

Fig. 13. Precision, recall and F-measure achieved by the three matching tools on 10 scenarios

Table 4. Average matching quality of the tools: COMA++, SF and YAM (single and tuned with 3 parameters)

	Precision	Recall	F-measure
COMA++	66 %	38 %	48 %
SF	61 %	43 %	50 %
YAM	81 %	65 %	71 %
YAM-recall	68 %	**78** %	73 %
YAM-domain-specific-scenarios	80 %	72 %	**76** %
YAM-expert-correspondences (5 %)	88 %	90 %	**89** %

Harmony schema matcher [35, 44] combines multiple matching algorithms by using a vote merger. The vote merging principle is a weighted average of the match scores provided by each match voter. A match voter provides a confidence score for each pair of schema elements to be matched. Then, the Similarity Flooding strategy [32] is applied to adjust the confidence scores based on structural information. Thus, positive confidence scores propagate throughout the graph. An interesting feature of Harmony lies in its graphical user interface for viewing and modifying the discovered schema correspondences through filters.

RiMOM [28] is a multiple strategy dynamic ontology matching system. Different matching strategies are applied to a specific type of ontology information. Based on the features of the ontologies to be matched, RiMOM selects the best strategy (or strategy combination) to apply. When loading the ontologies, the tool also computes three feature factors. The underlying idea is that if two ontologies share similar feature factors, then the strategies that use these factors should be given a high weight when computing similarity values. For instance, if the *label meaningful* factor is low, then the *Wordnet-based strategy* will not be used. Each strategy produces a set of correspondences, and all sets are finally aggregated using a linear interpolation method. A last strategy dealing with ontology structure is finally performed to confirm discovered correspondences and to deduce new ones. Contrary to other approaches, RiMOM does not rely on machine learning techniques to select the best strategy. It is quite similar to the *AHP* work (described hereafter) in selecting an appropriate matcher based on the input's features.

AgreementMaker [9] provides a combination strategy based on the linear interpolation of the similarity values. The weights can be either user assigned or evaluated through automatically-determined quality measures. The system allows for serial and parallel composition where, respectively, the output of one or more methods can be used as input to another one, or several methods can be used on the same input and then combined. The originality of AgreementMaker is the capability of manually tuning the quality of matches. Indeed, this tool includes a comprehensive user interface supporting both advanced visualization techniques and a control panel that drives the matching methods.

In [31], the authors propose a machine learning approach, SMB. It uses the Boosting algorithm to classify the similarity measures, divided into first line and second line matchers. The Boosting algorithm consists in iterating weak classifiers over the training set while re-adjusting the importance of elements in this training set. Thus, SMB automatically selects a pair of similarity measures as a matcher by focusing on harder training data. A specific feature of this algorithm is the important weight given to misclassified pairs during training. Although this approach makes use of several similarity measures, it mainly combines a similarity measure (first line matcher) with a decision maker (second line matcher). Empirical results show that the selection of a pair does not depend on their individual performance. Thus, only relying on one classifier is risky.

In a broader way, the STEM framework [26] identifies the most interesting training data set which is then used to combine matching strategies and tune several parameters such as thresholds. First, training data are generated, either manually (i.e., an expert labels the entity pairs) or automatically (at random, using static-active selection or active learning). Then, similarity values are computed using pairs in the training data set to build a similarity matrix between each pair and each similarity measure. Finally, the matching strategy is deduced from this matrix thanks to supervised learned algorithm. The output is a tuned matching strategy (how to combine similarity measures and tune their parameters). The framework enables a comparative study of various similarity measures (e.g., Trigrams, Jaccard) combined with different strategies (e.g., decision tree, linear regression) whose parameters are either manually or automatically tuned.

The MatchPlanner approach [18] makes use of decision trees to select the most appropriate similarity measures. This approach provides acceptable results with regard to other matching tools. However, the decision trees are manually built, thus requiring an expert intervention. Besides, decision trees are not always the best classifier, as shown in Sect. 5.

eTuner [27] aims at automatically tuning schema matching tools. It proceeds as follows: from a given schema, it derives many schemas which are semantically equivalent. The correspondences between the initial schema and its derivations are stored. Then, a given matching tool (e.g., COMA++ or Similarity Flooding) is applied to the set of correspondences until an optimal parameters configuration of the matching tool is found. eTuner strongly depends on the capabilities of the matching tool, and it has to be integrated in an existing matching tool by a programmer. Conversely, YAM learns a dedicated matcher according to a given matching scenario. It is also able to integrate important features like user preference between recall and precision. Contrary to eTuner, YAM is extensible in terms of similarity measures and classifiers, thus enhancing the capabilities of our tool.

Authors of [30] have proposed to select a relevant and suitable matcher for ontology matching. They have used Analytic Hierarchical Process (AHP) to fulfill this goal. They first define characteristics of the matching process divided into six categories (inputs, approach, usage, output, documentation and costs). Users then fill in a requirements questionnaire to set priorities for each defined characteristic. Finally, AHP is applied with these priorities and it outputs the most suitable matcher according to user requirements.

COMA/COMA++ [2,11] is a hybrid matching tool that incorporates many independent similarity measures. It can process Relational, XML, RDF schemas as well as ontologies. Internally it converts the input schemas as trees for structural matching. It provides a library of 17 element-level similarity measures. For linguistic matching it utilizes a user defined synonym and abbreviation tables, along with n-gram name matchers. Similarity values between each possible pair of elements and for each similarity measure are stored in a similarity matrix. Next, the combination of the values is performed using aggregation operators such as *max*, *min*, *average*. Different strategies, e.g., reuse-oriented matching or fragment-based matching, can be included, offering different results. For each source element, pairs with a combined similarity value higher than a threshold are displayed to the user for validation. COMA++ supports a number of other features like merging, saving and aggregating match results of two schemas.

Similarity Flooding (SF) and its successor Rondo [32,33] can be used with Relational, RDF and XML schemas. These input data sources are initially converted into labelled graphs and SF approach uses fix-point computation to determine correspondences between graph nodes. The algorithm has been implemented as a hybrid matcher, in combination with a terminological similarity measure. First, the prototype does an initial element-level terminological matching, and then feeds the computed candidate correspondences to the structural similarity measure for the propagation process. This structural measure includes a few rules, for instance one of them states that two nodes from different schemas are considered similar if their adjacent neighbours are similar. When similar elements are discovered, their similarity increases and it impacts adjacent elements by propagation. This process runs until there is no longer similarity increasing. Like most schema matchers, SF generates correspondences for pairs of elements having a similarity value above a certain threshold. The generation of an integrated schema is performed using Rondo's *merge* operator. Given two schemas and their correspondences, SF converts the schemas into graphs and it renames elements involved in a correspondence according to the priorities provided by the users.

Glue [13], and its predecessor LSD [12], are also based on machine learning techniques. They have four different learners, which exploit different information from the instances. The name learner (Whirl, a nearest-neighbour classifier) makes predictions using word frequency (TF/IDF distance) on the label of the schema elements. The content learner (also based on Whirl and TF/IDF) applies a similar strategy to the instances associated to each schema element. A Naive Bayes classifier considers labels and attributes as a set of tokens for performing text classification. The XML learner (based on Naive Bayes too) exploits the structure of the schema (hierarchy, constraints, etc.). Finally, a meta-learner, based on stacking, is applied to return a linear weighted combination of the four learners.

AUTOMATCH [5] is the predecessor of AUTOPLEX [4], which uses schema instance data and machine learning techniques to find possible correspondences between two schemas. An attribute dictionary contains attributes with a set of possible instances and their probability. This dictionary is populated using Naive Bayesian algorithm to extract relevant instances from Relational schemas fields. A first step consists of matching each schema element to dictionary attributes, thus computing a similarity value between them according to the number of common instances. Then, the similarity values of two schema elements that match the same dictionary attribute are summed and *minimum cost maximum flow* algorithm is applied to select the best correspondences. The major drawback of this work is the importance of the data instances. Although this approach is interesting on the machine learning aspect, that matching is not as robust since it only uses one similarity function based on a dictionary.

The main difference between YAM and all these matchers lies in the level of abstraction. Theoretically, YAM could generate most of these matching tools. This is actually the case with MatchPlanner [18]. The most relevant existing approach to YAM is the configuration tool eTuner, since both approaches discover the best configuration of a matcher. Yet, eTuner's capabilities are limited compared to YAM: it has to be plugged into an existing matching tool (which requires programming skills) and it totally depends on that matching tool, especially for the method which combines similarity measures. Thus, it does not offer the extensibility and self-tuning features encompassed in YAM.

7   Conclusion

In this paper, we have presented YAM, the first extensible and self-tuning factory of schema matchers. Instead of producing correspondences between schemas, YAM generates a dedicated schema matcher for a given matching scenario. This is made possible by formalizing the matching problem as a classification problem. In addition, we described how to integrate user requirements into the generation process so that the dedicated matcher fulfills the needs and preferences of the user. Our approach is also the first work to let users choose the promotion of either precision or recall. Experiments have shown that the dedicated matchers generated with YAM obtain acceptable quality results with regard to reputed matching tools. Finally, we outline here the lessons learned:

- We have demonstrated a strong need for a schema matcher factory;
- Our experiments support the idea that machine learning classifiers are suitable for the matching task and that the traditional aggregation functions are not always the most efficient method for combining similarity measures;
- We have studied the impact and the benefits on the matching quality when the user provides preferences such as the promotion of recall/precision or input expert correspondences.

In the future, we first plan to test further classifiers. Indeed, there exist a large number of machine learning classifiers of which we have experimented only a subset. Among them, the meta-classifiers base their predictions using the results of several classifiers and therefore offer the possibilities for improving matching quality. In a similar fashion, we foresee the possibility to deduce some correspondences between the matching results of all matchers. These highly probable correspondences could serve as input expert correspondences to produce a smart dedicated matcher. Finally semi-supervised learning [7] could be used to improve the accuracy of the dedicated matcher: the intuition is to include in the training data some unlabelled pairs from the schemas to be matched.

References

1. Altschul, S.F., Erickson, B.W.: Optimal sequence alignment using affine gap costs. Bull. Math. Biol. **48**(5–6), 603–616 (1986)
2. Aumueller, D., Do, H.-H., Massmann, S., Rahm, E.: Schema and ontology matching with coma++. In: SIGMOD, pp. 906–908 (2005)
3. Bellahsene, Z., Bonifati, A., Rahm, E. (eds.): Schema Matching and Mapping. Springer, Heidelberg (2011)
4. Berlin, J., Motro, A.: Autoplex: automated discovery of content for virtual databases. In: Batini, C., Giunchiglia, F., Giorgini, P., Mecella, M. (eds.) CoopIS 2001. LNCS, vol. 2172, pp. 108–122. Springer, Heidelberg (2001)
5. Berlin, J., Motro, A.: Database schema matching using machine learning with feature selection. In: Pidduck, A.B., Mylopoulos, J., Woo, C.C., Ozsu, M.T. (eds.) CAiSE 2002. LNCS, vol. 2348, p. 452. Springer, Heidelberg (2002)
6. Bernstein, P.A., Madhavan, J., Rahm, E.: Generic schema matching, ten years later. PVLDB **4**(11), 695–701 (2011)
7. Chapelle, O., Schölkopf, B., Zien, A. (eds.): Semi-supervised Learning. MIT Press, Cambridge (2006)
8. Cohen, W., Ravikumar, P., Fienberg, S.: A comparison of string distance metrics for name-matching tasks. In: Proceedings of the IJCAI 2003 (2003)
9. Cruz, I.F., Antonelli, F.P., Stroe, C.: AgreementMaker: efficient matching for large real-world schemas and ontologies. PVLDB **2**(2), 1586–1589 (2009)
10. Djeddi, W.E., Khadir, M.T.: Ontology alignment using artificial neural network for large-scale ontologies. Int. J. Metadata Semant. Ontol. **8**(1), 75–92 (2013)
11. Do, H.H., Rahm, E.: Coma - a system for flexible combination of schema matching approaches. In: VLDB, pp. 610–621 (2002)
12. Doan, A., Domingos, P., Halevy, A.Y.: Reconciling schemas of disparate data sources: a machine-learning approach. In: SIGMOD, pp. 509–520 (2001)
13. Doan, A.H., Madhavan, J., Dhamankar, R., Domingos, P., Halevy, A.Y.: Learning to match ontologies on the semantic web. VLDB J. **12**(4), 303–319 (2003)
14. Doan, A., Madhavan, J., Domingos, P., Halevy, A.: Ontology matching: a machine learning approach. In: Staab, S., Studer, R. (eds.) Handbook on Ontologies in Information Systems, pp. 397–416. Springer, Heidelberg (2004)
15. Dougherty, J., Kohavi, R., Sahami, M., et al.: Supervised and unsupervised discretization of continuous features. In: Proceedings of 12th International Conference on Machine Learning, vol. 12, 194–202 (1995)
16. Dragut, E., Lawrence, R.: Composing mappings between schemas using a reference ontology. In: Meersman, R. (ed.) OTM 2004. LNCS, vol. 3290, pp. 783–800. Springer, Heidelberg (2004)
17. Duchateau, F., Bellahsene, Z.: Designing a benchmark for the assessmentof schema matching tools. Open J. Databases (OJDB) **1**, 3–25 (2014). RonPub, Germany
18. Duchateau, F., Bellahsene, Z., Coletta, R.: A flexible approach for planning schema matching algorithms. In: Meersman, R., Tari, Z. (eds.) OTM 2008, Part I. LNCS, vol. 5331, pp. 249–264. Springer, Heidelberg (2008)
19. Duchateau, F., Bellahsene, Z., Roche, M.: A context-based measure for discovering approximate semantic matching between schema elements. In: Research Challenges in Information Science (RCIS) (2007)
20. Euzenat, J., Shvaiko, P.: Ontology Matching. Springer, Heidelberg (2007)
21. Fayyad, U.M., Irani, K.B.: On the handling of continuous-valued attributes in decision tree generation. Mach. Learn. **8**(1), 87–102 (1992)

22. Gal, A.: Uncertain Schema Matching. Synthesis Lectures on Data Management. Morgan & Claypool Publishers, San Rafael (2011)
23. Garner, S.R.: Weka: the waikato environment for knowledge analysis. In: Proceedings of the New Zealand Computer Science Research Students Conference, pp. 57–64 (1995)
24. Hammer, J., Stonebraker, M., Topsakal, O.: Thalia: test harness for the assessment of legacy information integration approaches. In: ICDE, pp. 485–486 (2005)
25. Hliaoutakis, A., Varelas, G., Voutsakis, E., Petrakis, E.G.M., Milios, E.: Information retrieval by semantic similarity. Int. J. Seman. Web Inf. Syst. **2**(3), 55–73 (2006)
26. Köpcke, H., Rahm, E.: Training selection for tuning entity matching. In: QDB/MUD, pp. 3–12 (2008)
27. Lee, Y., Sayyadian, M., Doan, A.H., Rosenthal, A.: eTuner: tuning schema matching software using synthetic scenarios. VLDB J. **16**(1), 97–122 (2007)
28. Li, J., Tang, J., Li, Y., Luo, Q.: Rimom: a dynamic multistrategy ontology alignment framework. IEEE Trans. Knowl. Data Eng. **21**(8), 1218–1232 (2009)
29. Lin, D.: An information-theoretic definition of similarity. In: ICML 1998, pp. 296–304 (1998)
30. Malgorzata, M., Anja, J., Jérôme, E.: Applying an analytic method for matching approach selection. In: CEUR Workshop Proceedings of Ontology Matching, vol. 225. CEUR-WS.org (2006)
31. Marie, A., Gal, A.: Boosting schema matchers. In: Meersman, R., Tari, Z. (eds.) OTM 2008, Part I. LNCS, vol. 5331, pp. 283–300. Springer, Heidelberg (2008)
32. Melnik, S., Garcia-Molina, H., Rahm, E.: Similarity flooding: aversatile graph matching algorithm and its application to schema matching. In: Proceedings of ICDE, pp. 117–128 (2002)
33. Melnik, S., Rahm, E., Bernstein, P.A.: Developing metadata-intensive applications with Rondo. J. Web Seman. **I**, 47–74 (2003)
34. Mitchell, T.: Machine Learning. McGraw-Hill Education, New York (1997). (ISE Editions)
35. Mork, P., Seligman, L., Rosenthal, A., Korb, J., Wolf, C.: The harmony integration workbench. J. Data Seman. **11**, 65–93 (2008)
36. Needleman, S., Wunsch, C.: A general method applicable to the search for similarities in the amino acid sequence of two proteins. J. Mol. Biol. **48**(3), 443–453 (1970)
37. University of Illinois: The UIUC web integration repository (2003). http://metaquerier.cs.uiuc.edu/repository
38. Paulheim, H., Hertling, S., Ritze, D.: Towards evaluating interactive ontology matching tools. In: Cimiano, P., Corcho, O., Presutti, V., Hollink, L., Rudolph, S. (eds.) ESWC 2013. LNCS, vol. 7882, pp. 31–45. Springer, Heidelberg (2013)
39. Peukert, E., Eberius, J., Rahm, E.: Rule-based construction of matching processes. In: Proceedings of the 20th ACM International Conference on Information and Knowledge Management, CIKM 2011, New York, pp. 2421–2424. ACM (2011)
40. Resnik, P.: Semantic similarity in a taxonomy: an information-based measure and its application to problems of ambiguity in natural language. J. Artif. Intell. Res. **11**, 95–130 (1999)
41. Secondstring (2014). http://secondstring.sourceforge.net/
42. Shvaiko, P., Euzenat, J.: A survey of schema-based matching approaches. In: Spaccapietra, S. (ed.) Journal on Data Semantics IV. LNCS, vol. 3730, pp. 146–171. Springer, Heidelberg (2005)

43. Shvaiko, P., Euzenat, J.: Ten challenges for ontology matching. In: Meersman, R., Tari, Z. (eds.) OTM 2008, Part II. LNCS, vol. 5332, pp. 1164–1182. Springer, Heidelberg (2008)
44. Smith, K., Morse, M., Mork, P., Li, M., Rosenthal, A., Allen, D., Seligman, L.: The role of schema matching in large enterprises. In: CIDR (2009)
45. Winkler, W.E.: String comparator metrics and enhanced decision rules in the fellegi-sunter model of record linkage. In: Proceedings of the Section on Survey Research, pp. 354–359 (1990)
46. Wu, X., Kumar, V., Quinlan, J.R., Ghosh, J., Yang, Q., Motoda, H., McLachlan, G.J., Ng, A., Liu, B., Philip, Y.S., et al.: Top 10 algorithms in data mining. Knowl. Inf. Syst. **14**(1), 1–37 (2008)
47. Xu, L., Embley, D.W.: Using domain ontologies to discover direct and indirect matches for schema elements, pp. 97–103 (2003)

Author Index

Printed in the United States
By Bookmasters